Boris Turko
Ivan Karbovnik
Oleksij Kushnir

ORGANIC LEDS: MATERIALS, TECHNOLOGIES AND ENGINEERING

Boris Turko
Ivan Karbovnik
Oleksij Kushnir

ORGANIC LEDS: MATERIALS, TECHNOLOGIES AND ENGINEERING

ScienciaScripts

Imprint

Any brand names and product names mentioned in this book are subject to trademark, brand or patent protection and are trademarks or registered trademarks of their respective holders. The use of brand names, product names, common names, trade names, product descriptions etc. even without a particular marking in this work is in no way to be construed to mean that such names may be regarded as unrestricted in respect of trademark and brand protection legislation and could thus be used by anyone.

Cover image: www.ingimage.com

This book is a translation from the original published under ISBN 978-620-6-79085-3.

Publisher:
Sciencia Scripts
is a trademark of
Dodo Books Indian Ocean Ltd. and OmniScriptum S.R.L publishing group

120 High Road, East Finchley, London, N2 9ED, United Kingdom
Str. Armeneasca 28/1, office 1, Chisinau MD-2012, Republic of Moldova, Europe
Printed at: see last page
ISBN: 978-620-6-65513-8

B. I. Turko, I. D. Karbovnyk, O. O. Kushnir, G. I. Klim,

B.S. Sadovyi, V.S. Vasiliev

ORGANIC LEDS:

MATERIALS, TECHNOLOGIES AND ENGINEERING

The monograph is devoted to theoretical and experimental studies of the optical and electrical properties of such functional materials for OLED technologies as: ITO, reduced graphene oxide, ZnO, Alq_3, dicyanomethylene pyran and its derivatives.

For scientists and engineers specializing in electronic materials science, nano-, microelectronics, and solid-state physics, as well as for teachers, graduate students, and students of relevant physical, technical, and natural sciences.

TABLE OF CONTENTS

INTRODUCTION

The development of organic light emitting diodes (OLEDs) poses serious challenges for scientists and technologists. A deep understanding of the physics of emitting and non-emitting processes is required to ensure the required characteristics of the final device, and the search for new efficient OLED materials requires the study of the relationship between their structure and properties. Significant progress has been made in this area in recent years, and one of the most pressing tasks today is to create diode structures characterized by polarized emission. Such elements can serve as light sources for a variety of applications, from backlighting liquid crystal displays and contrast OLED screens to stereoscopic image projection systems and biomedical applications [1, 2]. The absence of the need for an additional polarizer filter (currently, elements that absorb approximately 50% of the original light are used to provide polarization) allows for simplified device design and reduced manufacturing costs, while increasing energy efficiency, brightness, color contrast, and viewing angles.

At present, there are already OLED devices that emit light with linear or circular polarization - among the first developments are OLEDs with polarized luminescence and a dichroic ratio $E/E_{\perp \text{II}} = 2.4$ [3]. It should be noted that for practical use, the dichroic ratio must be greater than 40 ($E/E_{\perp \text{II}} > 40$).

Currently, scientists use several approaches and mechanisms to produce and optimize OLEDs with polarized light emission. For example, it has been shown that linearly polarized light emission can be obtained by incorporating uniaxially oriented materials (e.g., oligomers or liquid crystals) into the emitting layers using uniaxial alignment methods, such as mechanical tensile/shear/rubbing alignment, alignment on specific substrates, Langmuir-Blodgett deposition, or liquid crystal self-organization. Although these methods have made some progress, the insufficient degree of polarization is still a significant obstacle in practice. In addition, damage and contamination of the radiating layer inevitably degrade the polarization efficiency [1, 2].

Other approaches have shown that luminescence with a high degree of polarization (but low light output) can be obtained by using nanostructures such as photonic crystals or metal lattices. The disadvantage of these approaches is the predominant use of solid-state rigid substrates, which is not compatible with the current strategy of OLED technology development (OLED on flexible substrates) [1, 2].

The following problems of OLED devices remain unresolved: degradation of properties due to the influence of oxygen and water molecules, relatively low glass transition temperatures, low mobility of charge carriers due to the amorphous nature of organic molecules [4]. One of the promising ways to solve them is to use nanocomposites to create hybrid organic-inorganic LEDs. To optimize such hybrid systems with controllable optoelectronic properties, a deep understanding of the processes of electronic energy transfer between organic and inorganic subsystems is necessary and important [5].

In OLED structures, the cathode is usually an indium tin oxide (ITO) film with good optical transparency and high electrical conductivity. However, poor electron injection causes high turn-on voltage and low luminescence efficiency due to the mismatch between the work energy levels of the ITO output (4.5-4.8 eV) and the lowest unoccupied molecular orbital (LUMO) of most organic electronically conductive materials (2.7-3.2 eV) [6]. Therefore, it is important to improve the electron injection in OLED structures to reduce the on voltage and achieve efficient luminescence by introducing selected electron injection layers (EILs) between the ITO cathode and electron transfer layers (ETLs). To solve this problem, metal oxides with n-type conductivity, polarized polymers, alkali metal compounds, and thin metal layers have been used, among others [6]. The n-type metal oxides (TiO$_x$, ZrO$_2$, SnO$_2$, ZnO) are characterized by high transparency and controllable electrical properties [6]. Among these oxides, ZnO is a promising option due to its high electron mobility, but in the band energy diagram, the position of the energy level of the conduction band of ZnO (~4.1 eV) does not significantly coincide with the LUMO levels (2.7-3.2 eV) of most organic electronically conductive materials, and thus there is a high energy barrier to the injection of electrons from ZnO into ETLs [6], which leads to a deterioration in device performance. An additional negative factor is the dissociation of excitons caused by defective energy levels in ZnO [6], so many studies have focused on the introduction of intermediate layers to lower the barrier and realize high-efficiency electron injection. It has been shown that electron injection from the ZnO layer can be enhanced by modifying its surface with various intermediate layers of self-assembled dipole molecules, polyelectrolytes, alkali metal compounds, ionic liquids, and carbon nanotubes with n-type conductivity due to the formation of interfacial dipoles [6].

Many scientists have studied the effects of incorporating ZnO nanoparticles into EILs, ETLs, and hole-blocking layers (HBLs) of OLEDs. The authors of [7] created a diode with ZnO nanoparticles, which had a maximum efficiency of 16.5 cd/A, which is approximately three times higher than that of a diode without ZnO nanoparticles. The authors of [8] also achieved a threefold increase in luminescence efficiency by using triangular-shaped ZnO nanoparticles, creating OLED ITO/ZnO nanoparticles/MEH-PPV/V O$_{25}$ /Al. In [9], the lower switch-on voltages of OLEDs with ETL from ZnO nanoparticles were explained by the presence of the Auger recombination mechanism in nanoparticles. The authors of [10] also attributed the improvement of LED characteristics after the introduction of

5 nm ZnO nanoparticles into ETL and HBL OLEDs to the Auger effect in nanoparticles.

It is promising to manufacture OLED devices from inorganic and carbon nanostructures in order to increase the operating temperature ranges (by increasing the glass transition temperature of organic layers) due to cohesive interactions [11] and to replace expensive ITO electrodes with limited mechanical flexibility [12].

Thus, the monograph devoted to the development of technologies for obtaining high-quality thin films and nanostructures based on ZnO, ITO, reduced graphene oxide (rGO), aluminum tris(8-hydroxyquinoline) (Alq_3), dicyanomethylene pyran (DCM) and its derivatives, and the study of their structure features, surface morphology, recombination and energy transfer processes, optical and electrophysical properties, is relevant not only from the point of view of fundamental physics, but also for expanding the practical application of the above materials in microelectronics, optics, instrumentation, etc.

To date, quite a few publications have been devoted to OLED materials and technologies (Fig. 1), including review articles, e.g., [13- 17], books and collective monographs, e.g., [2, 4, 18-20], and the OLED screen market is growing every year and will continue to grow (Fig. 2).

Fig. 1. Diagram comparing the number of publications and patents found on the SciFinder.cas.org resource for the search word "OLED" for the period from 2000 to 2020 as of October 9, 2020 [13].

*Only the number of mobile phones is indicated

Fig. 2. The potential of the OLED screen market [21].

In Ukraine, in recent years, scientists have also been working on OLED issues, in particular in Kyiv (Taras Shevchenko National University of Kyiv: Professor, Doctor of Technical Sciences V. Amirkhanov; Professor, Doctor of Physical and Mathematical Sciences V. Yashchuk) and Lviv (Lviv

Polytechnic National University: Professor, Doctor of Technical Sciences P. Stakhira). At present, the number of domestic books, monographs, and review articles on OLEDs is insignificant [22-24]. In the book [25] and in the textbook [26], 22 and 5 pages are devoted to organic LEDs, respectively.

The monograph consists of an introduction, three chapters, conclusions, and a list of references.

The first section is devoted to the literature review of information on the general principles of operation of single-layer and multilayer organic light-emitting devices. The existing types of OLEDs and OLED panels, the classification of OLED devices by their emission layer are described. The advantages and disadvantages of OLED devices are presented.

The second section contains the results of studies of the optical and electrical properties of such functional materials for OLED technologies as: ITO, reduced graphene oxide, ZnO, Alq_3 , dicyanomethylene pyran and its derivatives.

The third chapter is devoted to the study of electroluminescence of OLED structures with ITO/Alq configurations ITO/Alq_3 :DCM-5 (10 wt. %)/Alq_3 /Al and ITO/Alq_3 :DCM-18 (10 wt. %)/Alq_3 /Al.

The monograph is based on the materials collected by the authors during ten years of research conducted by them with the staff of the Scientific, Technical and Educational Center for Low Temperature Research of Ivan Franko National University of Lviv. These results have been substantially supplemented and generalized using data from other domestic and foreign research groups studying similar scientific problems. To a large extent, this was made possible through personal contacts at international scientific conferences and internships at foreign research institutions.

It is worth noting that the authors' research results have been published in leading international and domestic journals. The high level of citation of these works convincingly demonstrates the relevance and interest of the world scientific community in the results of such research and ensures their international recognition.

The authors are sincerely grateful for the support of the Ministry of Education and Science of Ukraine and the National Research Foundation of Ukraine, 2020.02/0217 "Light-generating low-dimensional structures with polarized luminescence based on organic and inorganic materials".

SECTION 1

BASIC INFORMATION ABOUT OLED TECHNOLOGIES

1.1. General principles of operation of single-layer and multilayer organic light-emitting devices

Table 1.1 shows in chronological order the most important events in the development of OLED technology, in our subjective opinion.

Table 1.1

Chronology of OLED technology development [19, 20].

Year	Event.	Company/university
early 1950s	French scientist André Bernanose and his colleagues discovered electroluminescence in organic materials by applying a high-voltage alternating current to transparent thin films of acridine orange and quinacrine dye.	University of Nancy
1960	Martin Pope, who was born into a family of immigrants from Ukraine, and his colleagues first demonstrated the electroluminescence of anthracene crystals. However, despite intensive research, the first organic electroluminescent	New York University

11

Year	Event.	Company/university
	devices had significant drawbacks (high operating voltage, recrystallization, low emission brightness of 2.5 cd/m² and efficiency, and a short lifetime of tens of minutes).	
1987	Kodak employees Ching Wang Tang and Steven Van Slyke have created the first multilayer OLED.	Eastman Kodak
1990	Jeremy Barrows and his colleagues at the Cavendish Laboratory have created the first OLED based on conjugated polymers.	University of Cambridge
1991	Professor Junji Kido used lanthanide coordination compounds as an emission layer in OLEDs.	Yamagata University
1995	Dr. Peter Dyriklev and his colleagues in the Laboratory of Applied Physics have created the first OLED with polarized luminescence $E_\perp /E_{II} = 2.4$.	Linköping University
1999	Professor Stefan Forrest created OLEDs based on	University of Michigan

	phosphorescent iridium complexes.	
2014	A flexible OLED screen was manufactured.	LG display

The scheme for creating OLED devices with two layers of organic matter between the electrodes instead of one was first proposed by Kodak. According to it, everything is covered with glass covered with a thin ITO anode layer on the OLED side. Immediately adjacent to it is the first organic layer, 75 nm thick of aromatic diamine as a *p-type* semiconductor, followed by the main light-emitting layer of a film consisting of a compound belonging to the class of fluorescent metal chelate complexes. Many organic compounds based on the same aromatic carbohydrates can be used as *n-type* semiconductors. Finally, the last layer is the cathode, which consists of a mixture of magnesium and silver with an atomic ratio of 10:1. This system has a thickness of less than 500 nm, along with the backlight, which, among other things, is what it is. When a current of 2.5 V or more is applied, the base layer begins to emit photons, the flux of which becomes more intense as the current increases, amplifying almost linearly and allowing for a brightness of more than 1000 Cd per square meter at a voltage of less than 10 V. The peak intensity of the spectrum falls at a wavelength of 550 nm, which corresponds to the green color. Naturally, in addition to the obvious advantages, the scheme also had disadvantages. There was the durability, or rather the lack thereof - in the initial experiments, the luminosity (luminosity in physics is the luminous flux emitted by a unit area of a luminous surface (a surface that emits, reflects, or transmits light flux) in all directions (in the middle of the solid angle 2P)) at a constant voltage was halved after 100 hours of continuous operation, and problems with certain parts of the spectrum - in particular, with the blue. Nevertheless, the breakthrough was obvious, given that before that, a voltage of about 100 V was required to obtain more or less normal luminosity. A large number of companies joined in solving the remaining problems (today, about a hundred companies and universities are working on OLED), and most of them can already be considered solved to some extent. New OLED materials are much more complex combinations of substances than they were at the dawn of their history. New chemical formulas for the base layers, separate enrichment additives, each responsible for its own part of the spectrum - red, blue, green.

The main efforts of developers are now aimed at improving the characteristics of organic semiconductors. The successes are more than impressive: although the latest promising OLED materials remain the least durable in the blue spectrum, they still have a lifespan of up to 14,000 hours even under blue luminosity conditions. Red and green colors give up to 40,000, and universal white - 20,000 hours. This is already decent, considering that for the same digital cameras, for example, the average screen life is considered normal at 1000 hours. In addition, commercial products will obviously be based on the classic scheme used in liquid crystal displays (LCDs), when the screen consists of solid white OLED emitters with color filters responsible for giving color to specific pixels. In addition, new materials significantly improve the physical parameters of OLEDs. In particular, they raise the upper limit of the operating temperature range above 100°C [27].

Electroluminescence is the luminescence of a medium that occurs when an electric current is passed through it. It is a direct conversion of electrical energy into light energy.

The use of organic compounds for electroluminescence research began relatively recently, in the 1960s, when the electroluminescence of anthracene crystals was first demonstrated. However, despite intensive research, the first organic electroluminescent devices (OLEDs) had significant drawbacks, such as high operating voltage, recrystallization, low emission brightness (2.5 cd/m^2) and efficiency, and had a short lifetime (tens of minutes). The intensive development of research in the field of electroluminescent materials began in the 80s after the discovery of Ching Wang Tang and Stephen Van Slyke, who showed that by using a multilayer structure, it is possible to reduce the operating voltage and significantly increase the quantum efficiency of electroluminescence. They used Alq$_3$ as the active layer in the CRT [27].

Another very important discovery was made in 1990 by Jeremy Barrows and colleagues. They demonstrated the possibility of using π - conjugated polymers as emitting layers in CRTs. It was shown that the mechanism of electroluminescence of such materials is the injection of electrons and holes from the electrodes into the working layer, followed by their recombination to form photons [27].

Currently, single-layer, multilayer, planar and heterogeneous organic CRTs are known.

A typical OLED is a sandwich-like structure with a thickness of ~100 nm and consists of one or more layers in addition to electrodes (Fig. 1.1 a, b).

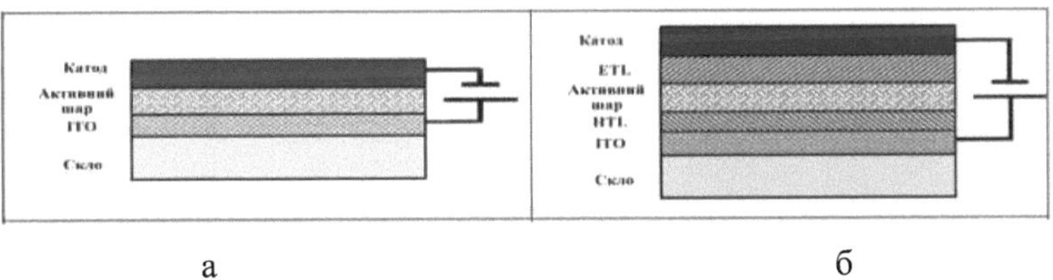

aб

Fig. 1.1. Structure of single-layer (a) and multilayer (b) OLED [27].

A single-layer OLED (Fig. 1.1a) consists of a glass substrate with a transparent electrode deposited on it, an active layer, and a metal cathode sputtered on top. To obtain high luminosity and efficiency, as well as to achieve low operating voltages, a substance that meets a number of requirements should be used as the active layer in such a scheme [27].

First, it must have both electronic and hole conductivity. That is, it must have a structure of energy levels similar to the structure of bands in semiconductors. In the case of organic compounds, this analogy is achieved by the structure of the highest occupied molecular orbital (HOMO) and the lowest unoccupied molecular orbital (LUMO). The energy interval between the HOMO and LUMO corresponds to the forbidden zone in the semiconductor. The energy interval between HOMO and LUMO should correspond to the energy of photons in the optical range of the spectrum [27].

Secondly, the structure of the substance must provide for the possibility of merging an electron and a hole to form an excited state, which can subsequently be highly likely to be illuminated. As a rule, in practice, the requirement of a high probability of illumination is equivalent to the requirement of a high quantum yield of luminescence of a substance [27].

Third, this substance must form homogeneous amorphous films. This is necessary for the proper operation of OLEDs, since a non-continuous film will lead to a short circuit between the electrodes, and a film that is heterogeneous in thickness or contains small crystals will lead to heterogeneity in current density, uneven luminosity, accelerated local processes of substance degradation with the formation of gas and electrode delamination, etc.

The main requirements for the anode are: good transparency in the visible spectral range, good conductivity, and correspondence of the output to the HOMO position of the active layer substance. For a large number of substances, a thin (~100 nm) ITO film meets these requirements most fully [27].

As a cathode in a single-layer OLED, a metal with an output work close to or slightly less than the LUMO depth of the active layer substance should be selected. For most substances suitable for the active layer, this depth is between 2 and 3 eV. All metals with work yields in this range (e.g., calcium, barium, ytterbium) are reactive. As a result, cathode layers made of these metals are sensitive to environmental influences and can react with the materials of organic layers, which causes instability in OLED performance and limits the reliability of the devices created. This greatly increases the level of technological requirements for the practical manufacture of devices. The use of materials that are resistant to environmental influences, such as Al and In, is limited due to the relatively high value of the output operation (approximately 4.3 eV), which leads to an increase in the operating voltage, a decrease in the quantum efficiency and lifetime of OLEDs [27].

The problem of creating materials with a low work output that could be used as a cathode is very relevant now. Currently, various alloys are widely used: Mg:Ag (10:1) and Al:Li (99:1), as well as combinations of different layers: LiF/Al, CsF/Al, LiF/Ca/Al [27].

Thus, the first, simplest and crudest, but important criterion for selecting materials for OLEDs is the correspondence of the electrode material yields to the HOMO and LUMO positions of the active layer substance. Therefore, the positions of these levels are usually shown in the OLED energy diagram (Fig. 1.2). In Fig. 1.2a shows the case of materials perfectly matched by the position of the energy levels. In such an OLED, as a rule, the values of the currents and quantum efficiency are determined by the mobility of carriers in the active layer and the value of the quantum yield of the luminescence of the working substance. The type of power schemes of such an OLED in the operating mode is shown in Fig. 1.3 a [27].

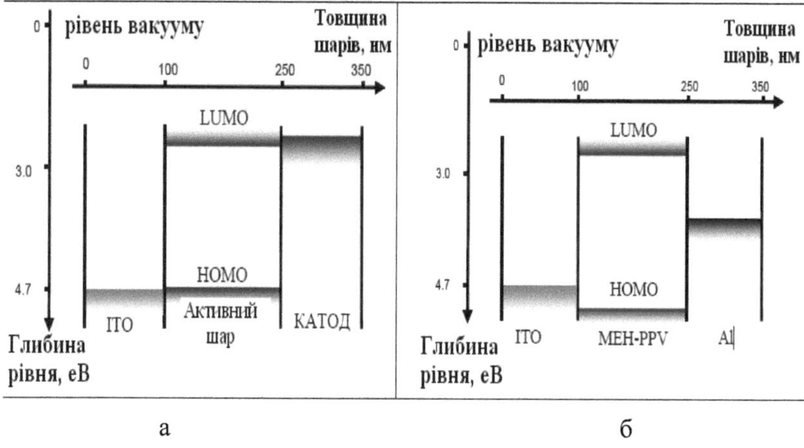

Fig. 1.2. Energy schemes of single-layer OLEDs: ideal energy scheme (a);
example of a suboptimal structure (b) [27].

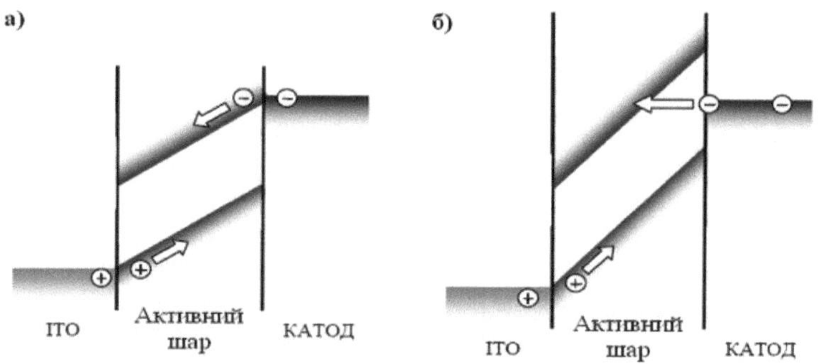

Fig. 1.3. Energy diagram of OLED in the operating mode (a) and with a large
barrier for electron injection in the operating mode (b) [27].

However, it is impossible to fully satisfy all of the above requirements in single-layer OLEDs. And the first difficulty is usually the selection of electrodes with output works corresponding to the HOMO and LUMO positions of the active layer substance [27].

An example of a single-layer OLED circuit with a large barrier to electron injection in the operating mode is shown in Fig. 1.3 б. In this

18

scheme, electrons must tunnel through this barrier to reach the LUMO (Fig. 1.3 b), which greatly reduces the electric current and increases the operating voltage of the device. In addition, the mobilities of electrons and holes differ greatly in most substances suitable for the active layer. These circumstances lead to the fact that the electron and hole currents flowing through the OLED are highly unbalanced. As a result, the quantum efficiency of OLEDs, i.e., the ratio of the number of emitted photons to the number of electrons passing through the structure from the cathode to the anode, is low, because it is determined in this case by the smallest of the carrier currents [27].

Additional layers used in multilayer OLEDs are designed to rid the device of this disadvantage (Fig. 1.4). The most commonly used layers are the so-called hole transporting layer (HTL) and electron transporting layer (ETL). Their purpose is, firstly, to create an intermediate "step" in the OLED energy scheme between the electrode and the active layer, which facilitates the injection of carriers in the case of a large barrier, and, secondly, to prevent the through flow of carriers through the structure, i.e., to improve the balance between electron and hole currents. The HTL is placed between the anode and the active layer (Fig. 1.4 a) [27].

As a material for HTL, a substance with a HOMO that occupies an intermediate (best, middle) position between the work output of the anode material and the HOMO depth of the working substance is chosen. The LUMO position of the HTL substance should be high, the mobility of holes should be high, and the mobility of electrons should be low. In addition, it is desirable that the distance between HOMO and LUMO is larger than that of the active layer substance. This will make it impossible to transfer excitation energy from the working substance to the HTL substance, and will also ensure that the condition of high LUMO position is automatically fulfilled [27].

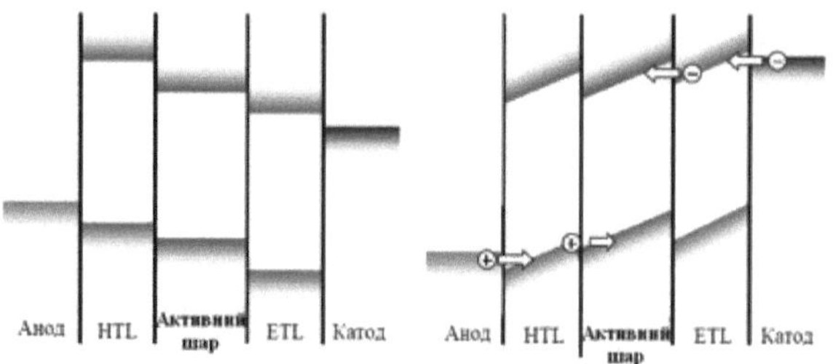

Fig. 1.4. Energy diagram of multilayer OLED layers (a) and multilayer OLED layers in operating mode (b) [27].

The ETL is placed between the active layer and the cathode, and has properties with respect to electrons similar to those of HTL with respect to holes (Fig. 1.4 a) [27].

Fig. 1.4 b shows the energy diagram of a multilayer OLED in the operating mode. It can be seen that the introduction of additional layers leads to the replacement of one high potential barrier with two low ones. Since the probability of tunneling, roughly speaking, decreases exponentially with the height of the barrier, the probability of sequential tunneling through two low barriers is much higher than through one high barrier, which leads to easier injection of carriers into the active medium. Examples of substances used as ETL and HTL are shown in Figs. 1.5 and 1.6, respectively [27].

Fig. 1.5. Chemical structure of molecules of materials used as ETLs in OLEDs [27].

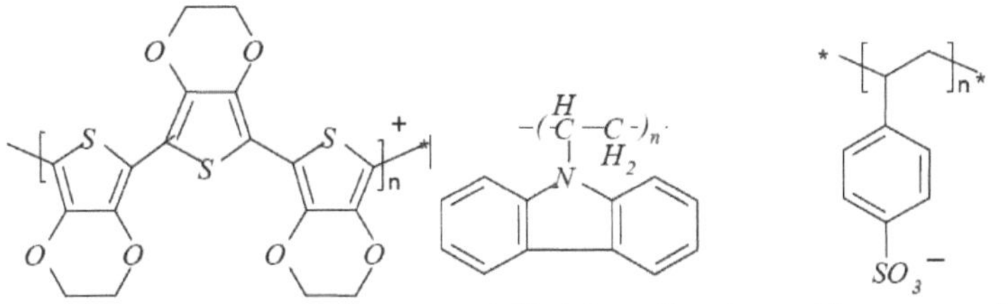

полі (етилендіоксітіофен) (PEDOT) полі (N-вінілкарбазол) (PVK) полі (стирен сульфонат) (PSS)

Fig. 1.6. Chemical structure of molecules of materials used as NTLs in OLEDs [27].

1.2. Types of OLED and OLED panels

1.2.1. OLED panels with passive and active matrices. There are OLED screens with passive (PMOLED) (Fig. 1.7) and active (AMOLED) (Fig. 1.8) matrices. In PMOLED, the organic layer is placed between perpendicularly crossed anode and cathode strips. The intersections of cathode and anode strips form pixels from which light is emitted. An external circuit delivers current to the selected anode and cathode strips, determining which pixels turn on and which remain off. The brightness of each pixel is proportional to the amount of applied current [28].

PMOLEDs are easy to manufacture, but they consume more power than other types of OLEDs, mainly due to the power required for the external circuitry. PMOLEDs are most efficient for displaying text and icons and are best suited for use in small screens with a diagonal of 2 to 3 inches in devices such as cell phones, mp3 players, and fitness bracelets. Even with external circuitry, passive matrix OLED displays consume less power than liquid crystal displays [28].

AMOLEDs have solid layers of cathode, organic molecules, and anode, but the anode layer is applied to an array of thin-film transistors (TFTs) that forms a matrix. The TFT array itself is the circuitry that determines which pixels are turned on to form the image [28].

AMOLEDs consume less power than PMOLEDs because the TFT array requires less power than external circuits, so they are efficient for large displays. AMOLEDs also have a higher refresh rate, which is suitable for displaying video. AMOLEDs are best used in computer monitors, large-screen TVs, and in electronic signage or billboards [28].

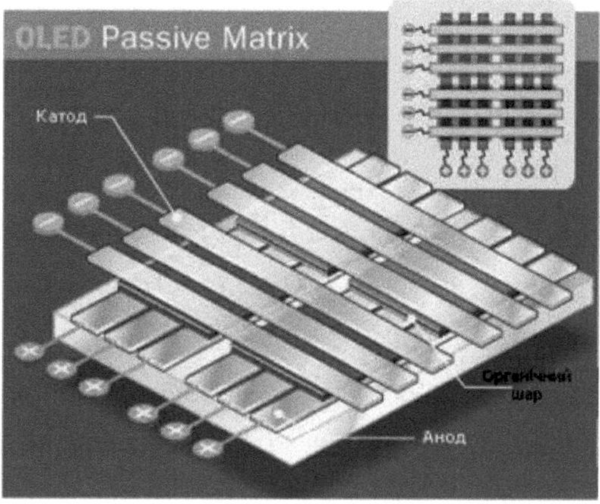

Fig. 1.7. OLED with a passive matrix [28].

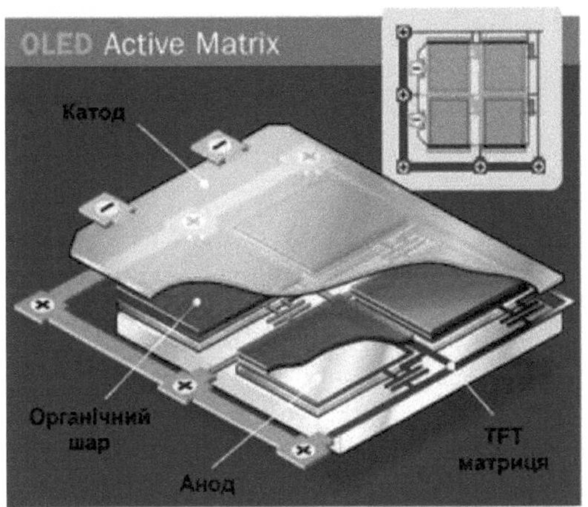

Fig. 1.8. OLED with an active matrix [28].

1.2.2. Transparent organic light-emitting diodes. A transparent OLED (TOLED) consists of only optically transparent components (substrate, cathode, and anode), and when turned off, it is 85% more transparent than its substrate. When a TOLED display is turned on, it transmits light in both directions. A TOLED display can be either active or passive. This technology can be used to make windshield displays (Fig. 1.9) [28].

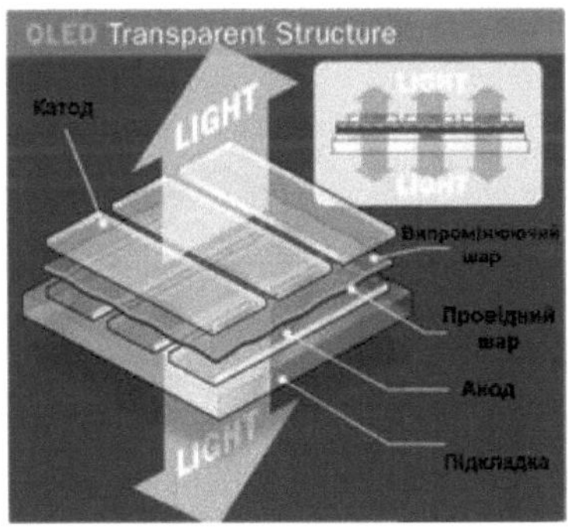

Fig. 1.9. Transparent OLED [28].

1.2.3 Organic LEDs with a radiating top surface. Organic LEDs with an emitting top surface have an optically opaque or reflective substrate. It is most appropriate to use them in displays together with an active matrix (Fig. 1.10) [28].

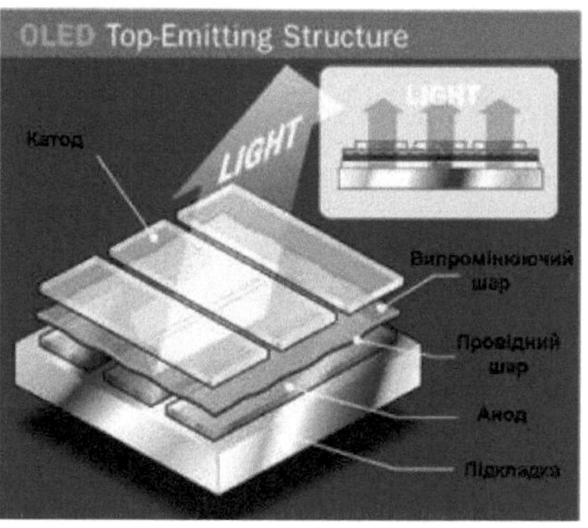

Fig. 1.10. Organic LED with a radiating top surface [28].

1.2.4. Flexible organic light-emitting diodes. A flexible OLED (FOLED) has a substrate of very flexible metal foil or plastic. FOLEDs are lightweight and transparent, and their use, for example, in mobile phones and pocket computers, can reduce the number of breakdowns of these devices. Potentially, flexible OLEDs can be attached to fabrics to create "smart" clothing (Fig. 1.11) [28, 29].

1.2.5. Organic white light-emitting diodes. The current state of development of organic electronics contributes to the active replacement of traditional power sources with lighting systems using organic light-emitting white light-emitting structures (WOLED) and the emergence of the latest displays based on WOLED technology on the market. Today, approaches to integrating WOLED structures into intelligent systems are also being actively developed. The feasibility of widespread commercialization of WOLED technology is driven by a number of factors, including low energy consumption and the absence of harmful materials in WOLED construction, which creates the preconditions for both global energy savings and a reduction in industrial environmental impact. In addition, the WOLED electroluminescence spectrum is as close as possible to the spectrum of solar radiation, which means that white light from WOLED is comfortably perceived. It should be noted that WOLED radiation is the result of the total electroluminescence of a two-color (blue-orange or blue-yellow) or three-color (superposition of blue, green and red) nature. Based on these approaches, WOLEDs are optimized using new design and technological solutions.

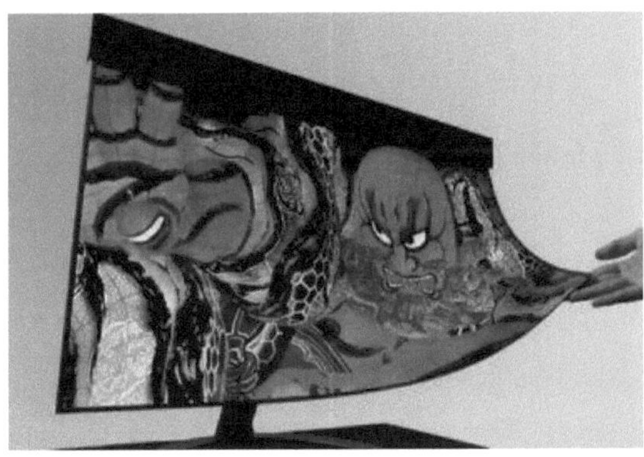

Fig. 1.11. Sharp and NHK (Japan Broadcasting Corporation) demonstrated on November 13, 2019, the world's first 30-inch flexible OLED screen with 4K resolution, weighing 100 g, 0.5 mm thick, and which can be twisted into a 4 cm diameter "tube"

With regard to energy efficiency and high quality of the white saturation color gamut, there are still problems to be overcome, including optimization of electron and hole injection, improvement of transport properties of current carriers, and efficiency of exciton generation and recombination in WOLED emitters (Fig. 1.12) [30].

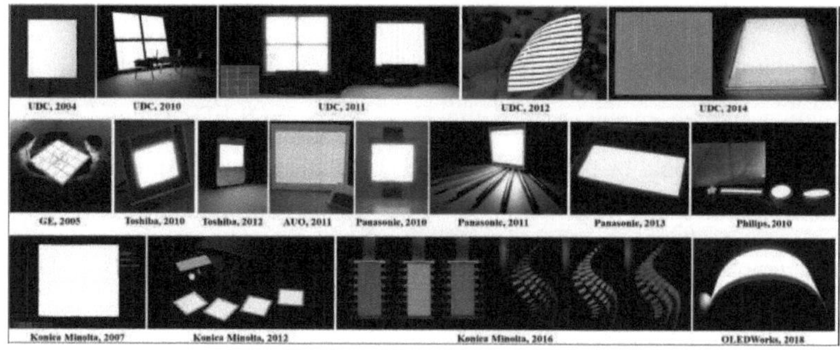

Fig. 1.12. WOLED lighting panels developed by commercial companies [30].

1.3. Gradation of OLED devices by their emission layer

There are many ways to classify OLED devices. One of them is the gradation of OLEDs into three generations according to their emission layer. This sorting is based on the mechanism of collecting excitons by the emitting medium (Fig. 1.13). The charge recombination in the emission layer corresponds to the spin statistics of 25% singlet and 75% triplet excitons [31]. First-generation OLEDs are manufactured with a purely organic fluorescent emitter, and therefore emit light only due to the presence of singlet excitons with a maximum possible internal quantum efficiency (IQE) of 25%. Second-generation OLEDs use heavy metal phosphorescent emitters that can harvest triplet excitons as well as the remaining 25% of singlet excitons through intersystem transitions (ISCs). Third-generation OLEDs have a thermally activated long-term fluorescence (TADF) material in their emitting layer, which can be either an organic or an organometallic compound [31]. TADF materials can utilize both singlet (S) and triplet (T) energy states through reverse intersystem transitions (RISC), which is a mechanism that converts triplet excitons into singlet ones through the use of thermal energy, without the need for rare and expensive elements. Thus, the

26

IQE for second- and third-generation OLEDs can theoretically reach 100% [31]. A small energy gap between the S and T energy levels is crucial for both ISC and RISC transmission mechanisms [31]. Organic TADF emitters have the combined advantages of first- and second-generation OLEDs, i.e., high stability and brightness, as well as a metal-free composition with low manufacturing costs. They can also consistently emit dark blue light, which has always been problematic for second-generation OLED devices [31].

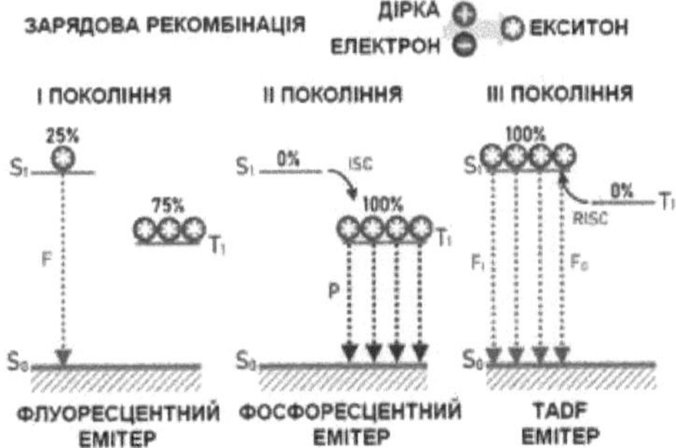

Fig. 1.13. Mechanisms of exciton accumulation in organic LEDs of the 1st, 2nd and 3rd generation [31].

1.4. Advantages and disadvantages of OLED devices

OLED devices have a number of advantages over LCD and LED devices, in particular [16, 28]:

- They are thinner, lighter, and more flexible.

- OLEDs are brighter than LEDs.

- Unlike LCD , OLED displays do not require a backlight. LCD devices work by selectively blocking areas of the backlight to create a visible image, while OLED displays generate light themselves. Because OLED displays do not require a backlight, they consume much less power than LCDs (most of the power used by LCDs goes to the backlight). This is especially important for battery-powered devices such as mobile phones.

- OLED -displays are easier to manufacture and can be produced in larger sizes.
- OLED displays have large viewing fields, approximately 170 degrees. Since LCDs work by blocking light, they cannot be viewed from certain angles. OLED devices emit their own light, so they have a much wider viewing range.
- OLED displays have a higher contrast and refresh rate, which minimizes motion blur.
- They can operate in a much wider range of operating temperatures than LCDs.

The disadvantages of OLED devices include [16, 28]:
- Short working life. While the red and green OLED layers are sufficient for 46,000 to 30,000 hours of operation, the blue layer is currently only able to function effectively for about 14,000 hours.
- OLED screens use three times more power to display an image with a white background. The use of a white background results in a shorter battery life for mobile devices.
- OLED screens are difficult to view in direct sunlight.
- Costly production.
- Water/moisture can easily damage the OLED.

SECTION 2
STUDY OF SOME FUNCTIONAL MATERIALS FOR OLED TECHNOLOGIES

2.1. Low-temperature technology for producing transparent ITO films with high conductivity

The term "transparent electronics" is considered to be born in 1997, when an article by Japanese researcher Kawazoe and his co-authors was published in the journal Nature. It described the production of a transparent highly conductive CuAlO oxide film$_2$ with *p-type* conductivity. The same issue of Nature published a paper by Thomas, which, based on the results of Kawazoe's work, discussed the prospect of creating so-called invisible electronic circuits that would provide new prospects for the use of oxide materials, which had previously been used only as passive elements of electronic circuits [32].

The main areas of application of transparent electronics are touchscreen displays, flexible displays, organic LEDs, electroluminescent emitters, thin-film photovoltaic cells, various electronic and optical coatings for other functional electronics devices [32].

ITO is a conductive oxide material that is now widely used in modern electronic and sensor devices. Despite the high price of this material (approximately $1,000 per 1 kg) and the available alternatives, it still retains its dominant position in the market of transparent conductive coatings [32].

Scientific interest in ITO films is consistently high. They are obtained by various methods, in particular, electron beam evaporation [33], thermal vacuum evaporation [34], direct current magnetron sputtering [35], pulsed laser deposition [35], high-frequency (HF) magnetron sputtering [36],

29

reactive sputtering [37], spray pyrolysis [38], sol-gel method [39], centrifugation [40], inkjet printing [41], chemical vacuum deposition [42], metal-organic chemical vacuum deposition [43], etc. RF magnetron sputtering is one of the best methods for manufacturing thin oxide films. The advantages of this method are its suitability for sputtering refractory materials, versatility, and relatively low cost. The effectiveness of RF magnetron sputtering is due to the high activity of the gas component molecules, which is stimulated by the action of RF plasma [44, 45].

Most of the available publications in this area are devoted to the study of the influence of technological parameters of production and subsequent processing on the optoelectrical properties of ITO films. In particular, it has already been established that it is possible to increase the electrical conductivity and optical transparency of ITO films by thermal annealing at temperatures above 200°C [46-49].

There are a number of heat-sensitive electronic devices with optically transparent electrodes. Therefore, at various stages of their manufacturing process, the temperature should not exceed 100°C. Obtaining competitive high-quality ITO films based on low-temperature technologies on an industrial scale is still a problem that is being solved, including using magnetron sputtering [50].

ITO films, as optically transparent heaters, can be used to quickly and reliably heat glass and plastic components (windshields, headlights, etc.) in various vehicles to overcome fogging and/or icing in harsh climates, in order to improve the efficiency of currently used solutions. For example, the windshields of modern cars are laminated with a polyvinyl butyral (PVB) polymer film that contains tungsten micro-wires as a heating element. However, this type of heating demonstrates insufficient transparency and homogeneity of heat distribution across the windshield [50]. The outer lenses in new automotive LED headlights are also subject to fogging or freezing due to moisture condensation occurring inside the lens. Typically, headlamp housings have air vents with filters to recirculate air and prevent condensation, the number and position of which must be frequently changed after the headlamp is manufactured to optimize air recirculation, which is a complex and expensive process [50]. This has become a challenging issue in terms of visibility and safety for many original equipment manufacturers. As in the case of windshields, micro-wires are also commonly used as heating elements, interfering with automotive radio or light detection and ranging (RADAR and LIDAR) systems. Optically transparent conductive coatings can improve the performance of conventional heating elements by demonstrating high heat capacity with rapid temperature control and low thermal inertia without compromising their optical transmittance [50].

Treatment with ultraviolet (UV) light is used to clean the surfaces of ITO films from carbon compounds and change the ratio in the number of surface atoms of In, Sn, and O, which leads to an increase in the electron yield of ITO [51-58]. Increasing the electron yield from ITO is useful because it reduces the interface barrier and increases the efficiency of hole injection from ITO into the organic layer, i.e., improves the performance of organic solar cells and organic LEDs [52-58].

The number of publications on comprehensive studies of the effect of low-temperature annealing [59] or light irradiation on the structural, optical, and electrical properties of ITO films [58, 60] is insignificant. Therefore, in this section, we publish the results of our research devoted to solving this scientific and technological problem.

We obtained ITO films by RF magnetron sputtering of a commercial target (99.99 % purity, Sigma-Aldrich) on optically transparent glass substrates for light wavelengths above 300 nm in an argon atmosphere at a working gas pressure of 0.1 Pa, RF generator power of 75 W, distance between the target and the substrate of 60 mm, magnetic field induction of

0.1 T, without heating the substrates. The sputtering time was 1 hour. According to the ellipsometric measurements, the thickness of the films was approximately 0.8 μm.

Some of the ITO films were annealed in air for 1 hour at different temperatures in a muffle electric furnace "SNOL-0.2/1250" (Umega Group AB, Utena, Lithuania). The other part of the samples was irradiated for different periods of time with light from a 125 W quartz mercury lamp DRT-125 located at a distance of 10 cm from the surface of thin film samples [61].

The surface morphology of the experimental samples was studied by an atomic force microscope (AFM) Solver P47-PRO.

Ex situ ellipsometry measurements were performed with an ellipsometer LEF-3M. The light source was a He-Ne laser ($\lambda = 632.8$ nm).

To study the spectra of light absorption and reflection in the UV and visible regions of the spectrum, a portable fiber optic spectrometer by Avantes BV (Apeldoorn, the Netherlands) "AvaSpec-ULS2048L-USB2-UA-RS" with an entrance slit of 25 μm and a diffraction grating of 300 pts/mm was used. Light detection in the spectrometer is performed by a 2048-pixel CCD matrix. Special software was used for automated computer control of the spectrometer and processing of the spectra. The halogen-deuterium light source AvaLight-DHc and the light reflection standard WS-2 (Avantes BV, Apeldoorn, the Netherlands) were used.

The resistivity of ITO films was measured by the four-probe method at room temperature using a Keithley 2401 electrometer (Keithley Instruments Inc., Cleveland, USA) [49, 62]. Silver thin-film electrodes deposited by thermal vacuum evaporation in a vacuum universal post VUP-5M (SELMI OJSC, Sumy, Ukraine) were used. The ohmic nature of the contacts of these electrodes with the sample is confirmed by the corresponding volt-ampere characteristic (Fig. 2.1).

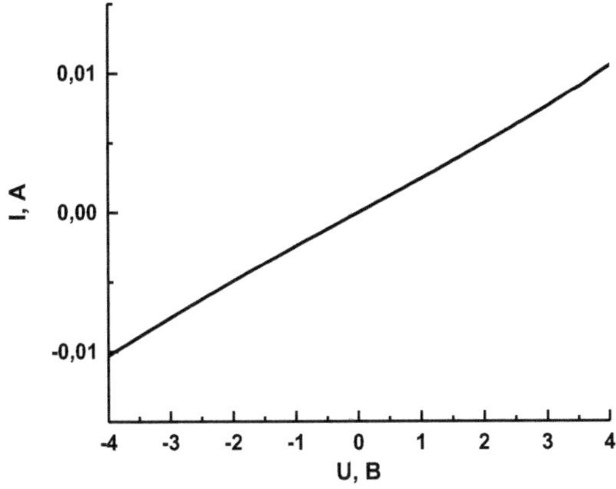

Fig. 2.1. Volt-ampere characteristic of annealed at 150°C thin film of ITO with silver contacts

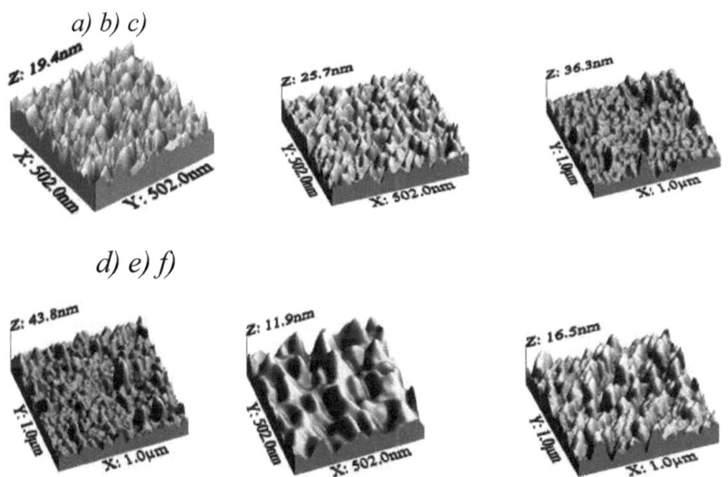

Fig. 2.2. Three-dimensional images of ITO films: a) without annealing and irradiation with light from a DRT-125 lamp; b) annealed at 50°C; c) annealed at 100°C; d) annealed at 150°C; e) after 5 minutes of irradiation with light from a DRT-125 lamp; f) after 20 minutes of irradiation with light from a DRT-125 lamp

Using the atomic force microscopy method, information on the surface morphology of ITO films was obtained (Fig. 2.2). In particular, such surface parameters as the root mean square roughness and the average size of crystallites were calculated (Table 2.1).

Table 2.1

RMS roughness and average size of crystallites in ITO films obtained under different technological conditions

Sample	Structure	Temperature. annealing, °C	Time irradiation with lamp light DRT-125, min	RMS roughness, nm	Average size of crystallites, nm
a	ITO/glass	-	-	1,5	5,1
б	ITO/glass	50	-	3,6	12,6
в	ITO/glass	100	-	3,4	13,4
г	ITO/glass	150	-	3,9	15,2
т	ITO/glass	-	5	1,8	5,5
д	ITO/glass	-	20	2,4	8,1

As the annealing temperature increases, the size of crystallites in the films increases. Illumination with a quartz mercury lamp also led to a structural rearrangement of ITO films. The irradiation slightly increased the size of crystallites, which is consistent with the data of [58, 60]. The authors of [58] irradiated ITO films with UV light with a wavelength of 365 nm from a 250 W source for 10 minutes, while in [60] ITO films were irradiated with light from a pulsed arc xenon lamp located at a distance of 22 mm from the sample. The irradiation duration was 20 s (20 cycles, with a short time interval between cycles for cooling, each cycle consisted of 8 pulses; the pulse duration was 200 µs, the interval between pulses was 75 µs, the intensity of the first pulse was 35 kW/cm^2 , the second - 26 kW/cm^2 , the third - 22 kW/cm^2 , the fourth - 18 kW/cm^2 , the fifth - 15 kW/cm^2 , the sixth - 12 kW/cm^2 , the seventh - 10 kW/cm^2 , the eighth - 8 kW/cm^2).

According to [60], untreated by annealing or irradiation, ITO films have a significant amorphous fraction consisting of visible light-absorbing compounds InO_x , SnO and atoms In^{+0} , Sn^{+0} . When irradiated, visible light is locally absorbed by metal atoms and partially oxidized compounds. Absorption excites these particles, and their subsequent vibrational relaxation leads to local heating of the place where they are located in the film [60]. This heat then leads to oxidation and crystallization reactions, in which small crystallites begin to grow at the nucleation sites. The optically opaque In^{+0} and InO_x inclusions crystallize into the optically transparent $In0_{23}$ oxide, just as the optically opaque Sn^{+0} and SnO inclusions crystallize into the optically transparent SnO_2 oxide, and the growth of crystallites stops when exposed to visible light. UV and infrared light will also lead to crystallite growth in ITO films, but at the same time, processes harmful to obtaining high-quality films, such as photodegradation and ablation, will occur. Photodegradation and ablation will reduce the electrical conductivity and optical transparency of ITO films [60].

Fig. 2.3 shows the absorption spectra of ITO films obtained under different technological conditions. According to these spectra, the lowest optical transparency in the visible range was observed for ITO films that were not annealed and not irradiated by the light of a DRT-125 lamp, and the highest for samples annealed at the highest temperature (150°C) and irradiated for 5 min by light from a DRT-125 lamp. A further increase in the irradiation time led to a decrease in the optical transparency of the films.

The absorption coefficients α were calculated using the Bouguer-

Lambert-Behr formula based on the transmission spectra. In general, the absorption coefficient is related to the bandgap width E_g by the relation [35, 37, 45, 46, 49]:

$$\alpha h\nu = B(h\nu - E_g)^r,\qquad(2.1)$$

where $r = 1/2$ for direct allowed transitions; B *is* a constant practically independent of photon energy.

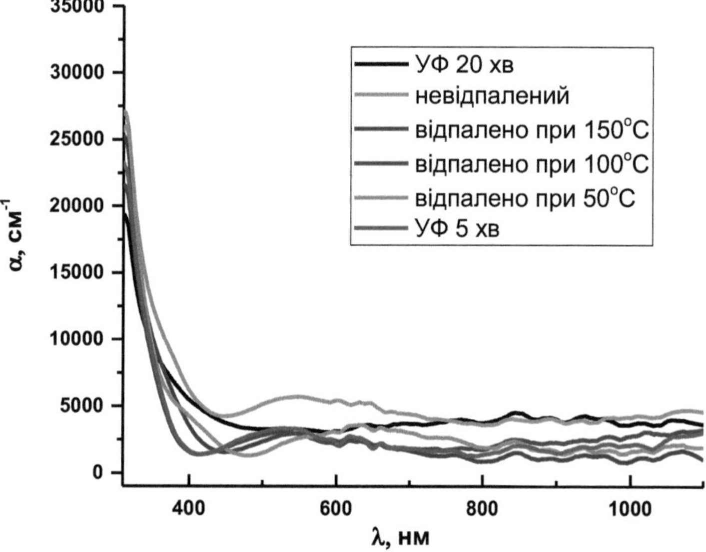

Fig. 2.3. Absorption spectra of ITO films obtained under different technological conditions: without annealing and irradiation with light from a DRT-125 lamp (green line); annealed at 50°C (orange line); annealed at 100°C (purple line); annealed at 150°C (blue line); after 5 min of irradiation with light from a DRT-125 lamp (red line); after 20 min of irradiation with light from a DRT-125 lamp (black line)

The values of the optical bandgap were obtained from the dependences of $(\alpha h v)^2$ on hv in the high absorption region with extrapolation of the linear sections of the curves to $(\alpha h v)^2 = 0$ [35, 37, 45, 46, 49].

Analysis of the fundamental absorption edge using Eq.1) shows that the optical band gap E_g : of the unannealed and unirradiated ITO film is approximately 3.59 eV, annealed at 50°C - 3.63 eV, annealed at 100°C - 3.65 eV, annealed at 150°C - 3.66 eV, irradiated for 5 min by the light of a DRT-125 lamp - 3.65 eV, irradiated for 20 min by the light of a DRT-125 lamp - 3.60 eV (Fig. 2.4). The obtained values of E_g correlate with the data of [37, 46, 63, 64].

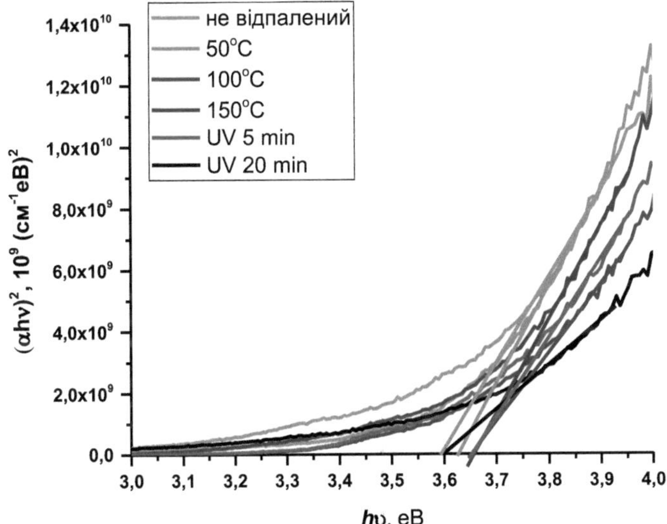

Fig. 2.4. Absorption spectra in the coordinates $(\alpha h v)^2 = f(hv)$ of ITO films: without annealing and irradiation with lamp light (green line); annealed at 50°C (orange line); annealed at 100°C (purple line); annealed at 150°C (blue line); after 5 min of irradiation with DRT-125 lamp light (red line); after 20 min of irradiation with DRT-125 lamp light (black line)

Fig. 2.5 shows the reflection spectra of ITO films obtained under different technological conditions. Thermal annealing and irradiation with light from a quartz mercury lamp reduce the values of the reflection coefficients in the visible wavelength range.

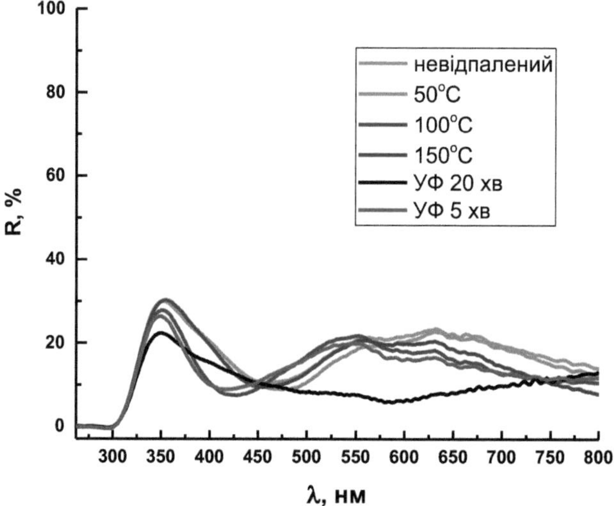

Fig. 2.5. Absorption spectra of ITO films obtained under different technological conditions: without annealing and light irradiation (green line); annealed at 50°C (orange line); annealed at 100°C (purple line); annealed at 150°C (blue line); after 5 min of irradiation with light from a DRT-125 lamp (red line); after 20 min of irradiation with light from a DRT-125 lamp (black line)

The values of the resistivity of ITO films obtained under different technological conditions are given in Table 2.2. With an increase in the annealing temperature to 150°C and an increase in the irradiation time with a quartz mercury lamp to 20 min, a decrease in the value of the resistivity was recorded, which is consistent with the data of [42, 46, 58, 60, 64-66].

Thus, we have studied the effect of low-temperature annealing and irradiation with light from a quartz mercury lamp DRT-125 on the structural, optical, and electrical properties of 0.8 μm thick ITO films deposited by the RF magnetron sputtering method on glass substrates that are optically transparent for light wavelengths greater than 300 nm. The substrates were not heated during deposition.

Table 2.2

Values of the electrical resistivity of ITO films obtained under different technological conditions

Structure	Temperature. annealing, °C	The time of exposure to the light of the DRT-125 lamp, min	Resistivity, Ohm-cm
ITO/glass	-	-	Non-conductive
ITO/glass	50	-	80
ITO/glass	100	-	6
ITO/glass	150	-	0,03
ITO/glass	-	5	556
ITO/glass	-	20	356

It has been found that low-temperature annealing for one hour and short-term irradiation with light from a quartz mercury lamp lead to an increase in the average size of crystallites and the root mean square surface roughness of ITO films. In particular, after annealing the ITO film at 150°C, the average crystallite size and RMS surface roughness increased approximately threefold, and after 20 minutes of irradiation - approximately 1.6 times, respectively.

The lowest optical transparency in the visible range of the spectrum among the studied experimental samples was observed in the unannealed and non-irradiated ITO films, and the highest in the annealed films at the highest temperature (150°C) and irradiated for 5 min with light from a DRT-125 lamp. An increase in the irradiation time leads to a decrease in the optical transparency of the films, which may be due to an increase in the number of defects due to the processes of ITO photodecomposition and ablation.

Thermal annealing and irradiation with light from a quartz mercury lamp reduce the reflection coefficient of ITO films in the visible spectrum.

Based on the analysis of the optical absorption spectra of ITO films, it was found that their optical band gap increases from 3.59 eV for an unannealed film to 3.66 eV with an increase in the annealing temperature to 150°C. In contrast, the dependence of the optical bandgap of ITO films on the time of irradiation with light from a quartz mercury lamp is non-trivial. Irradiation of ITO films with light from a quartz mercury lamp for 5 min leads to an increase in E_g from 3.59 eV to 3.65 eV, while after 20 min of irradiation E_g decreases to 3.6 eV. This is due to changes in the concentration of structural defects: its decrease after 5 min of light irradiation due to the growth of crystallite sizes and the formation of oxide compounds In O_{23} and SnO_2 , and, accordingly, its increase after 20 min of irradiation due to photodegradation and ablation processes.

With an increase in the annealing temperature to 150°C and an increase in the irradiation time with a quartz mercury lamp to 20 min, a decrease in the value of the resistivity of ITO films was recorded.

These results may be useful in the engineering of electronic devices based on materials with low thermal stability.

2.2. Absorption properties of reduced graphene oxide

It is well known that the performance of OLEDs is largely dependent on the injection of charge carriers from the electrodes [67]. ITO is the most widely used electrode for OLEDs because of its high optical transparency and good conductivity, but the cost of indium and its brittleness limit its use on flexible substrates. The electronic properties of the ITO surface are still far from ideal. In particular, there is a large energy barrier for hole injection at the ITO/organic interface due to the mismatch between the work energy level of the ITO yield and the LUMO level of the organic [6]. To improve the injection of holes and optimize their path to the radiative recombination zone, hole transfer layers are used in the emission layer. Poly-3,4-ethylene-dioxythiophene doped with polystyrene sulfonic acid (PEDOT:PSS) is widely used in polymer light-emitting diodes (PLEDs) as HTL. However, PEDOT:PSS can cause degradation of ITO due to its acidic nature, especially in the presence of moisture [68]. Transition metal oxides have also been used to reduce the energy barrier for hole injection, but vacuum deposition at high temperatures or annealing during synthesis from solutions can potentially harm the performance of OLED devices [69, 70]. Organic layers doped with phosphorus have been reported to effectively enhance hole injection in OLEDs, but the doping technology is complex [71]. Attempts have been made to use nanocarbon electrodes to improve hole injection with limited success [72-74]. A material such as graphene oxide (GO) can potentially replace PEDOT:PSS and ITO in flexible OLED devices.

Despite its great potential as a transparent conductor, the use of graphene as an anode in organic optoelectronics has been limited due to its relatively low electron yield and high surface resistance compared to ITO [75]. Graphene-based organic LEDs have demonstrated lower efficiency and higher operating voltages than ITO-based devices [76, 77]. High-efficiency

GO-based OLED devices can be fabricated using additional hole transfer layers [75]. The use of GO in OLEDs was reported in publications [78-82]. In [78], an OLED with GO as HTL demonstrated better performance compared to the reference device in terms of on voltage, current, and energy efficiency. In [79], an OLED with a reduced graphene oxide (rGO) anode was fabricated, but the performance of the device could not compete with that of ITO devices. rGO, due to its adjustable electron yield, was used as a cathode in inverted PLEDs and light-emitting electrochemical cells [80, 81]. In [82], OLED with GO as HTL demonstrated significantly better performance compared to the reference device with ITO.

The great interest in graphene and compounds based on it, in particular, GO and rGO, is primarily due to such unique physicochemical properties as high electrical and thermal conductivity, and the dependence of electrical characteristics on the presence of radicals of different nature on the surface. These properties are extremely important for the possible application of graphene as a basis for new nanomaterials with improved mechanical, electrical, and thermal physical characteristics, as well as an element of nanoelectronic devices [83].

Graphene, one of the variants of crystalline graphite, is a carbon nanomaterial, a layer of carbon atoms, one atom thick, connected in a two-dimensional crystal lattice of regular hexagons with a side of 0.142 nm, the vertices of which are occupied by carbon atoms. Graphene is very strong and ductile. Its uniqueness lies in its ability to exhibit conductor and semiconductor properties. The thermal conductivity of graphene is in the range of 3500 $W \cdot m^{-1} \cdot K^{-1}$ and 5500 $W \cdot m^{-1} \cdot K^{-1}$, which is a record value for all known materials [83].

GO is a kind of carbon compound that is a derivative of non-stoichiometric graphite oxide. It is a wrinkled two-dimensional carbon sheet with various oxygen-containing functional groups on its base plane and periphery, approximately 1 nm thick and with cross-sectional dimensions varying from several nanometers to several microns. By reducing GO, oxidized functional groups are removed and rGO is formed (Fig. 2.6) [83]. The study of rGO is currently a priority due to the manifestation of similar characteristics to graphene obtained by mechanical methods [83].

Among the various methods, the chemical reduction of GO to rGO is unique and attractive due to its ability to produce single-layer graphene on an industrial scale at relatively low cost [83]. In addition, rGO is easy to process and can be manufactured or self-assembled into macroscopic composite materials with controlled microstructures for use in various industries [83].

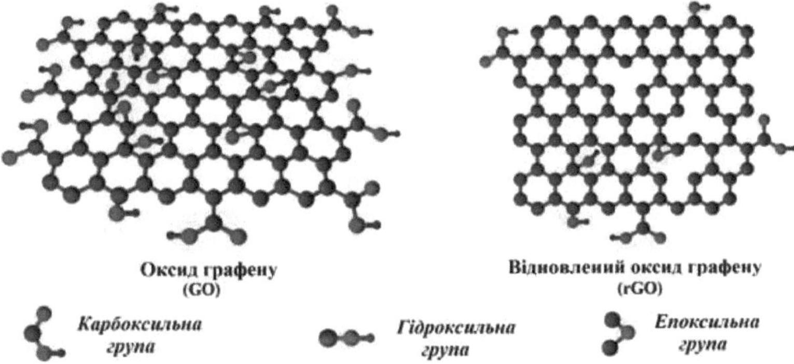

Fig. 2.6. Representation of the GO and rGO structure [83].

There are also products of covalent modification of GO, which have been developed primarily to solve the problem of agglomeration of GO in various solvents and further improve its dispersion and stability, as well as to ensure compatibility with various other objects [83].

The method of fabricating highly conductive and transparent nanoscale rGO bilayers (double layers) described in detail in [84] is promising for use in the production of organic LEDs. Therefore, at the initial stage, before fabricating OLEDs with rGO bilayers, we decided to investigate the absorption properties of the rGO bilayers deposited on quartz substrates provided to us by the authors of [84].

The rGO bilayers were obtained by depositing GO sheets on the surface of a quartz substrate, encapsulating them in a dense molecular polymer solution, and thermal reduction at 1100°C in nitrogen [84].

The rGO bilayer films had the following characteristics: thickness ~ 3.5 nm, sheet resistance 1.370 kΩ/sq, conductivity 2300 S/cm, and optical transparency ~ 90%. In Fig. 2.7 and Fig. 2.8 show the morphology of the synthesized rGO bilayers [61, 83].

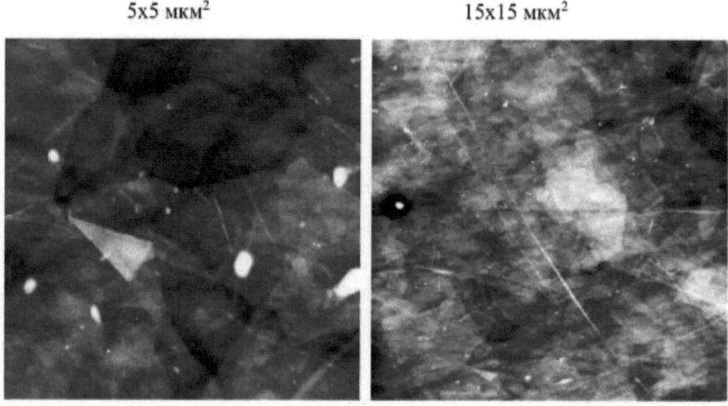

Fig. 2. 7. AFM image of an rGO bilayer on a quartz substrate. Vertical scale: 60 and 10 nm for 15 × 15 μm² and 5 × 5 μm² , respectively [61, 83].

Fig. 2. 8. Images obtained by a scanning electron microscope-microanalyzer REMMA-102-02 (SELMI OJSC, Sumy, Ukraine) of a two-layer rGO film deposited on a quartz substrate [61, 83].

To estimate the absorption of rGO in the visible spectral range, the minimum electronic transition energy was determined. The optical absorption spectrum of the bilayer rGO was measured at room temperature using an optical spectrometer. The absorption spectrum of the bilayer rGO contained two distinct maxima (Fig. 2.9) [61, 83]. The first of them (at 247 nm) corresponds to the electronic $\pi\pi^*$ transition (collective excitation) in graphite-like fragments of sp^2 -hybridized carbon [61, 83]. The position of this maximum, which is also called the π-plasmon or Van Hove feature, depends on the oxidation depth of the starting gases and ranges from 225-240 nm for deeply oxidized starting gases and up to 270 nm for unoxidized graphene. The shoulder at 290-320 nm corresponds to *nπ transitions* involving unshared electron pairs of oxygen atoms of carbonyl groups [61, 83]. In addition, the sample absorbs in the visible range. The absorption of the rGO layer in this range is associated with electronic transitions in the sp^2 -C clusters, the energy of which decreases with increasing cluster size [61,

83]. To estimate the minimum electronic transition energy E_{et}, an analysis was performed for the region $\lambda > 400$ nm (Fig. 2.10) [61, 83].

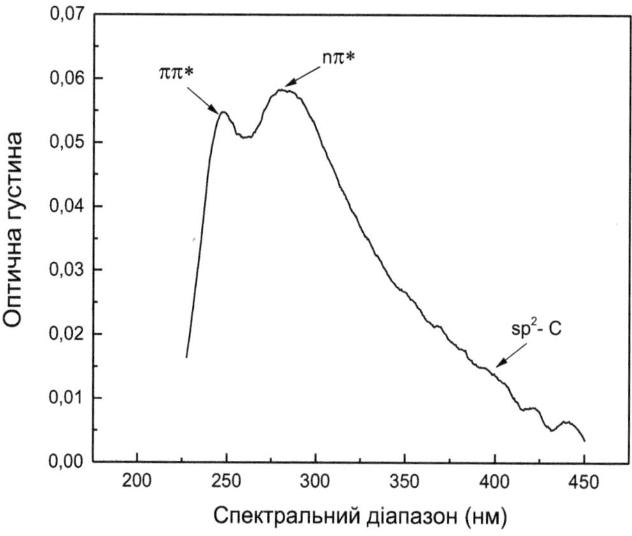

Fig. 2. 9. Absorption spectrum of the rGO bilayer [61, 83].

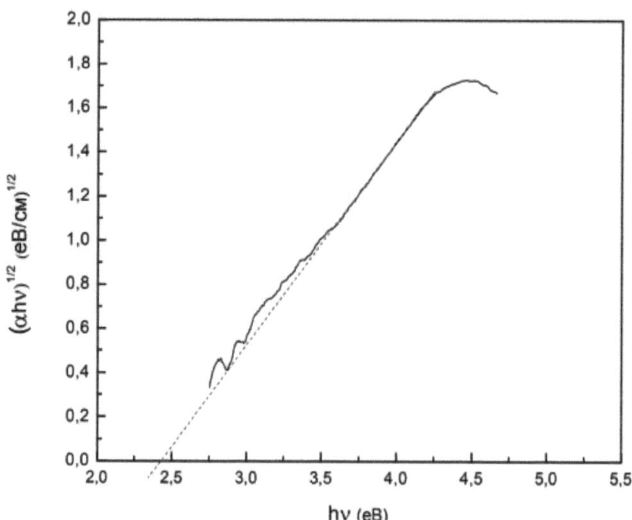

46

Fig. 2. 10. Absorption spectrum of the rGO bilayer in the coordinates

$(\alpha h v)^{1/2} = f(h v)$ [61, 83].

The Taus equation (2.1) was used to determine the type of transition under study [61, 83]. In our case, it was found that the absorption in the visible region of the spectrum is associated with indirect electronic transitions [61, 83]. Using the above relation, we determined E_{et} by extrapolating the linear part of the dependence of E on $(\alpha E)^{1/2}$ to the abscissa axis, as shown in Fig. 2.10. The obtained value $E_{et}\, E = 2.4$ eV [61, 83] is in agreement with the data given in other publications [85-87].

2.3. Low-temperature studies of the absorption spectra of a thin film of Alq3

Alq3 is very stable and is widely used in organic LEDs, solar cells, and organic field-effect transistors as an ETL or light-emitting layer material [88-90]. Alq3 is thermally stable, has a high glass transition temperature (172°C), and due to its polymorphic phase behavior, it is easily deposited as amorphous continuous thin films by vacuum thermal evaporation [90]. Important factors that determine the set of optical and emission characteristics of a material are the shape and position of its own absorption edge [45, 91-93]. For the optimal use of Alq3 films, it is necessary to have in-depth knowledge of their structure, optical and electrical properties.

There are a number of publications, for example [94-97], in which the authors describe the optical absorption of Alq3 thin films at room temperature. The low-temperature luminescence of these films has also been studied [98-102]. However, very little is known about low-temperature studies of the absorption spectra of Alq3 . Only one publication is known

[99], which presents the absorption spectra of Alq$_3$ thin film measured at room temperature and liquid nitrogen temperature.

Therefore, we have studied the temperature evolution of the absorption spectra of a thin film of Alq$_3$ in the temperature range of 16-320 K.

A thin film of Alq$_3$ with a thickness of 100 nm on a quartz substrate was obtained by thermal vacuum evaporation of Alq$_3$ powder (99.995 % purity, Sigma-Aldrich Corporation) from a quartz crucible at a pressure of 10^{-4} Pa. The thickness of the organic film was monitored online using a quartz film thickness gauge KITP-5 (Akademprylad, Sumy, Ukraine).

To study the spectra of light absorption in the UV and visible regions of the spectrum, a portable fiber optic spectrometer AvaSpec-ULS2048L-USB2-UA-RS (Avantes BV, Apeldoorn, the Netherlands) with an entrance slit of 25 µm, a diffraction grating of 300 pts/mm, and a resolution of 1.2 nm was used. Special software was used for automated computer control of the spectrometer and processing of the spectra. An AvaLight-DHc halogen-deuterium light source (Avantes BV, Apeldoorn, the Netherlands) was used. The sample was placed in a closed-circuit helium cryostat equipped with a DE-202A cryocooler (Advanced Research Systems, Macungie, USA) and a Cryocon 32 thermostat (Cryogenic Control Systems Inc., Rancho Santa Fe, USA).

Fig. 2.11 shows the absorption spectrum of a thin film of Alq$_3$ obtained at room temperature. The spectrum contains two broad absorption peaks with maxima at 265 nm and 392 nm and a hidden peak at approximately 340 nm. Our experimental results are consistent with the data reported in [96, 99, 103-108]. The nature of the high-energy band at 265 nm is associated with the presence of $p\pi$ * transitions, while the low-energy band at 392 nm is associated with $\pi\pi$ * transitions [96, 106]. According to [99], a small band at about 340 nm exists due to transitions to the lowest excited singlet states.

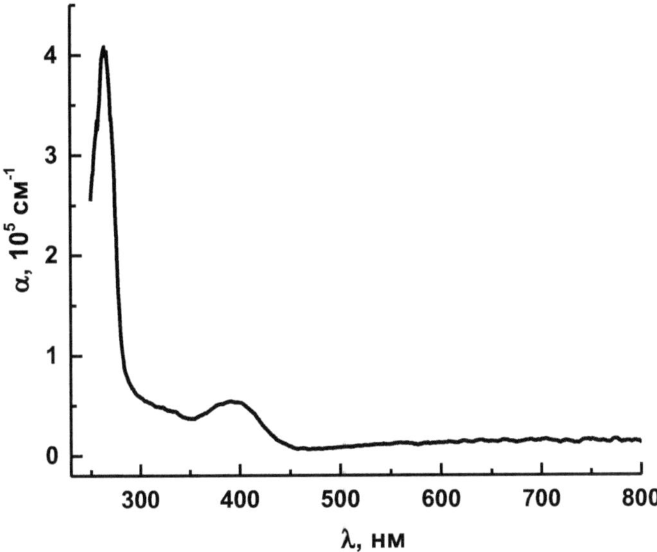

Fig. 2.11. Absorption spectrum in the UV and visible wavelength ranges of light from a thin film of Alq$_3$ obtained at room temperature

The absorption spectra of the Alq$_3$ film obtained at temperatures of 16 K, 70 K, 160 K, 280 K, and 320 K are shown in Fig. As can be seen from Fig. 2.12, the maximum value of the absorption coefficients was obtained at a temperature of 70 K. Cooling the sample from 150 K to 16 K leads to a very slight modification of the absorption spectral characteristics. The temperature evolution of the absorption spectra in our case correlates with the temperature evolution of the photo- and electroluminescence spectra described in [100, 101, 109-112]. According to the data presented in [110], at high excitation densities, the photoluminescence (PL) efficiency decreases with a decrease in temperature from 170 K to 15 K, and at low densities, the PL efficiency is practically independent of temperature. The observed quenching of FL at high excitation densities was attributed to singlet-singlet annihilation. The decrease in exciton annihilation with increasing temperature was attributed to the thermally activated population of unquenched localized exciton states. Above 190 K, the efficiency of FL began to decrease regardless of the excitation level, which was explained by the thermally activated release of excitons with subsequent migration to radiation-free centers outside the traps [110].

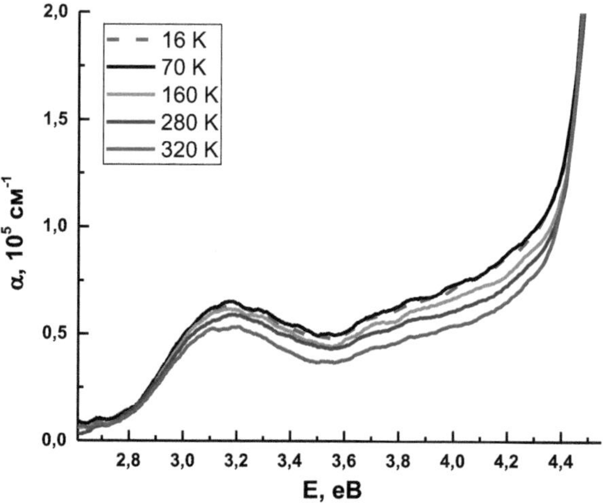

Fig. 2.12. Absorption spectra of Alq₃ thin film obtained at temperatures of 16 K, 70 K, 160 K, 280 K, and 320 K

According to [113], in non-excitonic amorphous semiconductors, a clear definition of the bandgap is often difficult due to the large static disorder that causes subband expansion. Organic semiconductors that are excitonic and partially amorphous exhibit even more complex subband features, including intermolecular hybrid states with charge transfer in technologically important mixtures of electron donors and acceptors, excitonic features, and trap states [113]. For Alq₃ , the dependence of the absorption coefficient α on the energy of the incident photon $h\nu$ can be written in the form of the Taus equation (2.1) [94-96, 106, 108, 109].

On the basis of Eq. (2.1) in the coordinates $\ln(\alpha h\nu)/\ln(h\nu)$, the type of electronic transitions for the absorption bands with maxima at approximately 265 nm and 392 nm was determined. The slopes of the obtained linear dependences were used to calculate the power factor r. In our case, the value of r was 0.5 for both absorption bands of Alq_3 .

Fig. 2.13 shows the absorption spectra measured at 16 K, 70 K, 160 K, 280 K, and 320 K in the region of the absorption edge at 392 nm in the coordinates $(\alpha h\nu)^2 = f(h\nu)$. It turned out that the value of the optical energy band gap of the Alq_3 thin film in this case practically does not change in the temperature range from 16 K to 320 K and is approximately equal to 2.86 eV. This value is in good agreement with the data for Alq_3 thin films (2.86 eV) reported in [96, 106, 114] and is slightly higher than for ε-*phase* Alq_3 crystals (approximately 2.82 eV) [109] or amorphous mer-Alq_3 (approximately 2.7 eV) [109].

Using Eq. (2.1) and Fig. 2.14, for temperatures of 16 K, 70 K, 160 K, 280 K, and 320 K, we obtained the corresponding values of the optical high-energy band gap of the Alq thin film$_3$: 4.48, 4.49, 4.47, 4.44, 4.43 eV. For comparison, at room temperature, a value of 4.44 eV was obtained for this energy gap in [94].

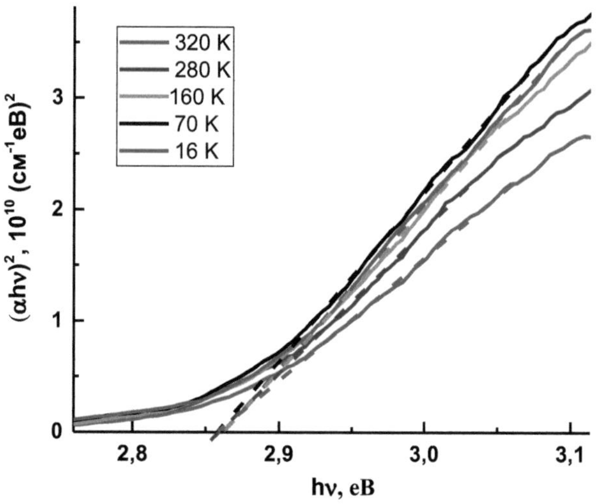

Fig. 2.13. Absorption spectra of Alq thin film₃ measured at 16 K, 70 K, 160 K, 280 K and 320 K in the absorption edge region at 392 nm in the coordinates $(\alpha h\nu)^2 = f(h\nu)$

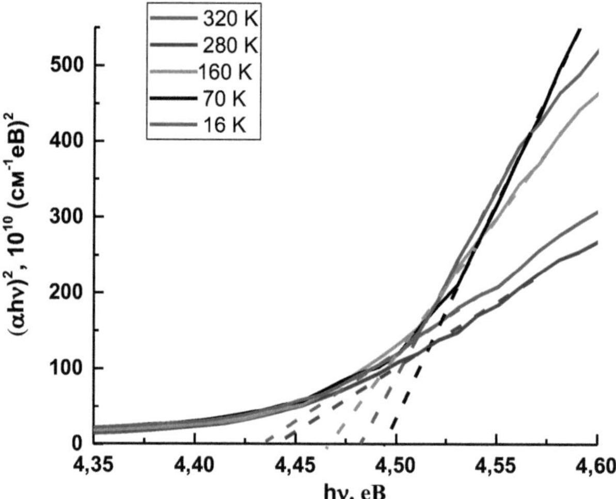

Fig. 2.14. Absorption spectra of Alq thin film₃ measured at 16 K, 70 K, 160 K, 280 K and 320 K in the region of the absorption edge at 265 nm in the coordinates $(\alpha h\nu)^2 = f(h\nu)$

Thus, for the first time, we have studied the absorption properties of Alq3 thin film in the UV and visible wavelengths in the temperature range of 16-320 K. The Taus equation was used to analyze the optical data. We calculated the values of two optical energy gaps for the Alq3 film at temperatures of 16 K, 70 K, 160 K, 280 K, and 320 K. It was found that the value of the first optical energy gap practically does not change in the temperature range from 16 K to 320 K and is approximately equal to 2.86 eV. While the following values were obtained for the second one: 4.48 eV, 4.49 eV, 4.47 eV, 4.44 eV, and 4.43 eV, respectively.

2.4. Manipulation of the refractive index of thin films Alq3 , DCM and DCM derivatives

The indisputable advantages of organic materials are usually considered to be their low cost, plasticity, high efficiency, and the variability of synthesis methods. One of the most widely studied organic compounds is Alq3 , DCM and its derivatives. The interest in these structures is due to the extremely wide range of their practical applications: they are used as components of active laser media [115, 116], OLEDs [89, 90, 115-117], sensors (explosives, in particular) [116-118], logic gates [116], dye-sensitized solar cells [119], etc.

Dicyanomethylene pyran is a well-known red dye and one of the first compounds discovered by Kodak employees in the process of manufacturing OLEDs and organic lasers. The DCM compound itself has a very low luminescence intensity due to significant intermolecular interactions that quench luminescence, so a guest-host system such as Alq3 :DCM is used [120, 121]. The distance between the LUMO and HOMO levels in Alq3 is larger than in DCM. Therefore, in the Alq3 :DCM compound, the excitation is radiation-free transferred from the Alq$_3$ molecule to the DCM molecule and then the radiative transition occurs. $_3$For high light yields, it is chosen to add 5-10 wt. % DCM. Concentrations higher than 10% lead to concentration quenching. There is also a concentration dependence of the location of the maximum emission band in Alq$_3$:DCM. Since the Alq$_3$ molecule is a polar molecule like DCM, a dipole-dipole interaction occurs in the composite and the phenomenon of solvatochromism is present. To prevent the formation of aggregates and reduce intermolecular interactions based on the organic compound DCM, almost immediately after its discovery, work began on the synthesis of new compounds with a more dendritic (tree-like) structure of molecules [120, 121].

It is known that in OLED devices, only a small portion of the generated light goes outside [122]. It turned out that it is possible to increase the quantum efficiency of OLEDs by about 30 % by reducing the refractive index of thin film layers in OLEDs [122]. The authors of [122, 123] used the oblique angle deposition method to reduce the refractive index of Alq thin films$_3$.

In this book, we present the results of ellipsometric studies of thin films obtained by thermal vacuum evaporation of Alq$_3$, dicyanomethylene pyran and its derivatives (DCM-5 and DCM-18) on glass substrates placed perpendicularly and at an angle of 10° to the vapor flow of the deposited substance.

Thin films of DCM, DCM-5, DCM-18 (C H N$_{252133}$) O, Alq$_3$:DCM (10 wt. % DCM) and Alq$_3$:DCM-5 (10 wt. % DCM-5) on glass substrates placed perpendicularly and at an angle of 10° to the vapor flow of the deposited substance were obtained by thermal vacuum evaporation of powders of starting organic compounds from quartz crucibles at a pressure of 10^{-4} Pa. DCM powder (99.1 % purity)- 4-(dicyanomethylene)-2-(4-amylcyclohexyl)-6[4-(dimethylaminoethyl)]-4H-pyran, was purchased from

Acros Organics (Geel, Belgium). Alq$_3$ powder with a purity of more than 98 % was purchased from Tokyo Chemical Industry Co., Ltd. The dicyanomethylene pyran derivatives DCM-5 and DCM-18 were synthesized in the Materials and Technologies of LCD Devices laboratory and kindly provided for research by leading researcher Oleksandr Kukhta. The thickness of the organic films was monitored online by a quartz film thickness gauge KITP-5 (Akademprylad, Sumy, Ukraine).

The resulting films were examined on an ellipsometer of the LEF-3M type (Fig. 2.15). The accuracy of the refractive index determination was± 0.005.

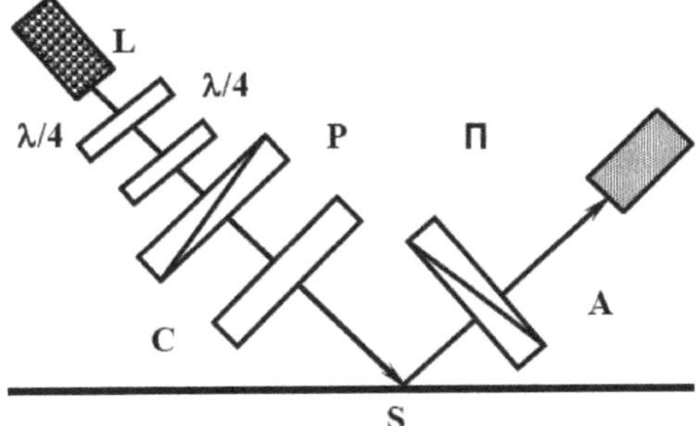

Fig. 2.15. Schematic of an ellipsometer: L - helium-neon laser ($\lambda = 632.8$ nm); $\lambda/2$ - plate that converts plane-polarized light into light with circular polarization; C - compensator; A - analyzer; P - light receiver (FEP); S - film on the substrate

Ellipsometry is a branch of optics that studies the change in the polarization characteristics of radiation after transmission or reflection at the interface between two media. The method consists in measuring the parameters of the polarization ellipse of the light reflected from the surface under study and comparing them with the corresponding parameters of the incident light [124].

The basic equation of ellipsometry establishes the relationship between microscopic (surface structure) and macroscopic (thickness and refractive index) characteristics of a sample and the ellipsometric parameters of the surface:

$$R_P / R_S = tg\Psi e^{i\Delta}, \quad (2.2)$$

where R_P, R_S are the Fresnel coefficients for the P- and S-components of the electromagnetic wave; Ψ and Δ- are the main parameters of the polarization ellipse of the reflected wave.

Parameters Ψ and Δ at given angles of incidence of light on the sample and wavelength of the radiation used are surface characteristics and are determined by the nature of the substance that makes up the sample, as well as the structure of the surface layer, surface quality (average roughness, presence of structural disturbances caused by polishing, etc.), presence of any film of a certain thickness, and properties of the environment in which the sample is located.

The value $\rho = tg\ e\ \Psi^{i\Delta}$ is called the relative reflection coefficient of polarized radiation. The specific analytical representation of the coefficients through the macroscopic and microscopic characteristics of the system under study, and, accordingly, the specific form of the last equation, depends on the choice of a particular surface model.

The basic equation of ellipsometry allows us to calculate the desired parameters of the system under study from the measured angles Ψ and Δ within the selected model (for example, to determine the thickness and refractive index of a film on a substrate with known characteristics- - the classical problem of ellipsometry).

The Drude equation (the basic ellipsometry equation within the single-layer model), which expresses the functional relationship of the ellipsometric parameters of the sample surface Ψ, $\varDelta$ measured under the specified experimental conditions with the five parameters of the system under study (refractive indices n_0 and absorption k_0 of the substrate, film n_1 and k_1, and film thickness d), is as follows:

$$tg\Psi e^{i\Delta} = \frac{R_{21P} + R_{10P}e^{-2i\delta}}{1 + R_{21P}R_{10P}e^{-2i\delta}} \frac{1 + R_{21S}R_{10S}e^{-2i\delta}}{R_{21S} + R_{10S}e^{-2i\delta}}, \quad (2.3)$$

where $2\delta=4\square(d/\square)n_1 \cos\theta_1$; $\square$ is the -wavelength; θ_1 is the refractive index; R_{10P}, R_{10S}, R_{21P}, R_{21S} are the Fresnel coefficients for the air-media-film and film-substrate interfaces.

The Drude equation is most often used to determine the thickness and refractive index of a transparent film on a substrate with a known refractive index.

Fig. 2.16 shows the chemical structure of the organic molecules that make up the studied thin films. A characteristic feature of DCM, which determines its functional properties, is the simultaneous presence of donor and acceptor groups in the molecule, connected by an unsaturated bridge, which leads to a noticeable separation of charges of different signs already in the ground state, which increases when the molecule is excited [124]. The DCM molecule is almost flat, but the aliphatic part of the molecule creates some steric hindrance to tight packing and prevents aggregation [126].

DCM

DCM-5

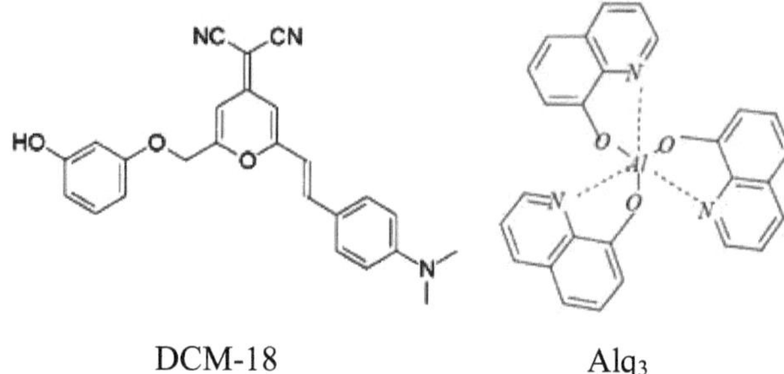

Fig. 2.16. Chemical structure of the molecules that make up the studied thin-film organic compounds

Table 2.3 shows the values of the refractive indices of DCM, DCM-5, DCM-18, Alq₃ :DCM (10 wt. % DCM), and Alq₃ :DCM-5 (10 wt. % DCM-5) thin films obtained by the ellipsometric method for a light wavelength of $\lambda = 632.8$ nm.

Table 2.3

The refractive indices (for $\lambda = 632.8$ nm) of organic thin films obtained on glass substrates placed perpendicularly $n_\perp$ and at an angle of 10° $n_\angle$ to the vapor flow of the deposited substance.

Structure and thickness of films	$n,_\perp \pm 0.005$	$n,_\angle \pm 0.005$
DCM, ~ 125 nm	2,230	2,178
DCM-5,~ 35 nm	2,377	2,201
DCM-18,~ 40 nm	1,805	1,729
Alq₃ :DCM (10 wt. % DCM),~ 35 nm	1,847	1,693
Alq₃ :DCM-5 (10 wt. % DCM-5),~ 30 nm	1,672	1,668

By using the method of thermal vacuum deposition at an oblique angle (10° to the vapor flow), the authors of [122] reduced the refractive index of

Alq$_3$ at a light wavelength of $\lambda = 525$ nm from 1.75 to 1.45 (Δ n = 0.3). If we compare our results for DCM, DCM-5, DCM-18, Alq$_3$:DCM (10 wt. % DCM), and Alq$_3$:DCM-5 (10 wt. % DCM-5) with the data reported in [122, 123] for Alq$_3$, in our case there are not such significant differences between the refractive indices of the films deposited on substrates placed perpendicularly and at an angle of 10° to the vapor flow. This fact can be explained by the fact that, unlike the authors of [122, 123], we dealt not with small molecules but with high molecular weight organic compounds. The molecule Alq$_3$ is a "small molecule". The term "small molecule" is derived from the fact that the carbon backbone of the molecule consists of only a short sequence and/or polycyclic aromatic compounds [127]. It is known that the adsorption of an evaporated substance on a substrate located at an oblique angle to the vapor flow results in the formation of islands that shade part of the substrate and prevent the vapor flow from entering the shaded areas, which leads to an ordered, porous, and columnar film growth [89, 122, 126-128]. And large molecules are more difficult to organize.

Thus, we have obtained thin films of DCM, DCM-5, DCM-18, Alq$_3$:DCM (10 wt% DCM) and Alq$_3$:DCM-5 (10 wt% DCM-5) by thermal vacuum evaporation on glass substrates placed perpendicularly and at an angle of 10° to the vapor flow of the deposited substance. The results of ellipsometric studies of these films are presented. It has been established that using the method of deposition of DCM, DCM-5, DCM-18, Alq$_3$:DCM (10 wt. % DCM) and Alq$_3$:DCM-5 (10 wt. % DCM-5) films at an oblique angle, it is possible to manipulate the values of their refractive indices at a light wavelength of $\lambda = 632.8$ nm in the range from n$\Delta \approx 0.05$ to n$\Delta \approx 0.18$. The results obtained will be useful in optimizing and designing high-performance organic LEDs from the studied thin films.

2.5. Polarized photoluminescence of thin films of Alq$_3$

Organic light-emitting diodes are being thoroughly researched and commercialized. One of the most urgent tasks in OLED technology is to obtain diode structures with a polarized nature of the emitted light and a dichroic ratio of more than 40 [2]. Polarized light has a wide range of applications, such as anti-reflective and 3D displays, encryption, optical communications, stereoscopic projection systems, and biomedicine [1, 2, 129].

Alq$_3$ is widely used in OLEDs [19]. In [122], it was shown that the refractive index of Alq$_3$ could be reduced from 1.75 to 1.45 by oblique angle deposition (OAD) of 10°, and then a low refractive index Alq$_3$ layer could be used in OLEDs and a 30% increase in efficiency could be achieved compared to the control device. The properties of polarized luminescence were not studied in [122]. Prior to that, [130] reported the production of photoluminescence with circular polarization of Alq$_3$ films deposited at a gliding angle (GLAD).

In this chapter, we report on the fabrication, characterization, and polarized luminescence of Alq$_3$ thin films on glass substrates, which we obtained by oblique angle deposition.

Organic layers of Alq$_3$ with a thickness of less than 50 nm were thermally deposited in a vacuum at a pressure of 10^{-4} Pa on two glass substrates placed perpendicularly and at an angle of 10° to the vapor flow of the deposited substance, respectively. Alq$_3$ powder with a purity of more than 98 % was purchased from Tokyo Chemical Industry Co., Ltd. The thickness of the organic films was monitored online using a quartz film thickness gauge KIT (Akademprylad, Sumy, Ukraine).

X-ray diffraction (XRD) measurements were performed by a STOE STADI P diffractometer with a position-sensitive linear detector in the Bragg-Brentano transmission geometry (Cu $K_{\alpha 1}$ radiation with $\lambda = 0.15406$

nm, Ge (111) monochromator, detector scan step: $0.480° 2\theta$, accumulation time: 320 s, angular resolution 2θ: 0.015°, 2θ-range: 5°-65°).

The Alq_3 powder was studied by differential thermal analysis (DTA) using a Linseis STA PT 1600 synchronous thermal analyzer (Linseis Messgeraete GmbH, Sielb, Germany). Heating from 298 K to 683 K at a rate of 10 K/min was carried out in a dynamic argon atmosphere.

The surface morphology of the experimental samples was studied by an atomic force microscope Solver P47-PRO (NT-MDT Co., Moscow, Russia).

The FTIR spectra at room temperature were measured using a quartz polarizer (Glan-Taylor prism) and a portable fiber optic spectrometer from Avantes BV (Apeldoorn, the Netherlands) "AvaSpec-ULS2048L-USB2-UA-RS" with an entrance slit of 25 μm, a diffraction grating of 300 ppm and a resolution of 1.2 nm. The signal accumulation time was 200 ms.

A rotating polarizer was placed between the photoluminescent source and the light detector. Special software was used for automated computer control of the spectrometer and spectrum processing. The samples were excited by the unpolarized light of an M365FP1 fiber-optic-coupled LED (λ = 365 nm, half-width of the diffraction line of maximum (FWHM) - 9 nm, LED output power - 15.5 mW, Thorlabs, Inc.)

To date, it is known that the Alq_3 molecule has two different geometric isomers: meridional (*mer-*) and facial (*fac-*) [109]. To the best of our knowledge, five crystalline phases of Alq_3 have already been identified: α-, β-, γ-, δ-, and ε- [109]. α- and β- have two mer-Alq_3 molecules per unit cell, γ- has two fac-Alq_3 molecules per unit cell, δ- has four fac-Alq_3 per unit cell, while ε- has three mer-Alq_3 per unit cell [109, 131, 132].

The X-ray diffractogram of the original Alq_3 powder is shown in Fig. 2.17. X-ray crystallographic analysis of OLED materials, including Alq_3 , is often quite difficult because OLED crystals are sometimes disordered and contaminated with other polymorphs.

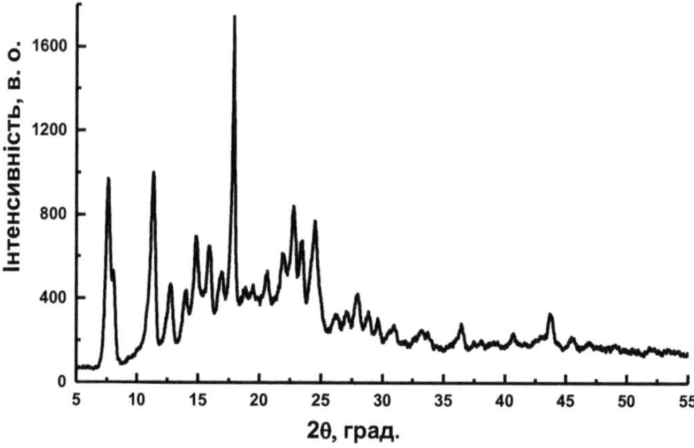

Fig. 2.17. X-ray diffractogram of the initial Alq powder3 [89].

To better understand the thermal factors associated with the generation of Alq phases3 , we used DTA. In Fig. 2.18 shows the DTA curve of the initial Alq_3 powder (initial phase -Alqε3). The phase transitions resulting from the heat treatment of Alq_3 are obvious. The phase transition temperatures are 623 K (from -Alqε3 to α-Alq_3) and 662 K (from α-Alq_3 to γ-Alq).3

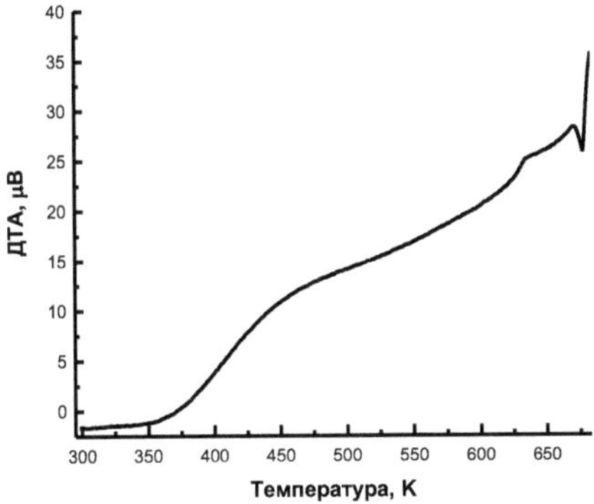

Fig. 2.18. DTA curve of the initial Alq powder₃ [89].

AFM micrographs of Alq thin films₃ are shown in Fig. 2.19.

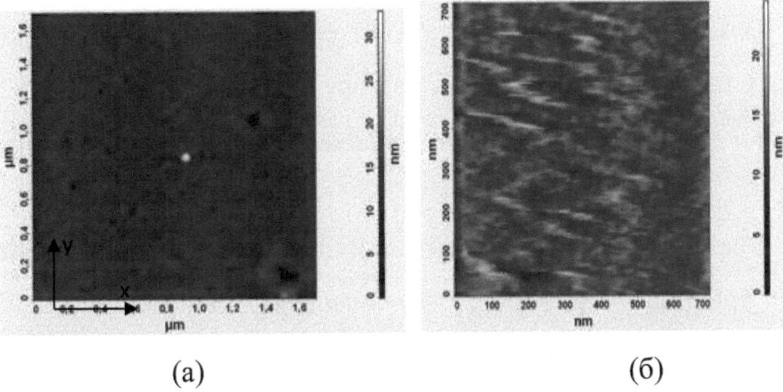

(a)   (б)

Fig. 2.19. AFM images of Alq₃ thin films on glass substrates placed at the deposition angle: a) - 90°, b) - 10° [89].

The FT-IR spectrum of Alq₃ powder obtained at room temperature showed a characteristic green emission with a broad band in the range from about 430 nm to 650 nm (Fig. 2.20).

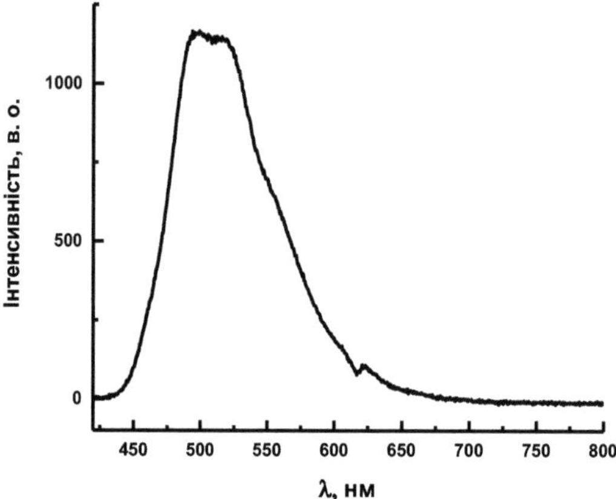

Fig. 2.20. FTIR spectrum of Alq₃ powder obtained at room temperature [89].

Based on the comparison of our experimental results obtained by DTA, X-ray and fluorescence spectroscopy with the literature [98, 102, 133-136], it was found that the powder contained crystalline phases ε -Alq₃ andα -Alq₃ .

The room-temperature polarized Raman spectra of Alq₃ thin films are shown in Fig. 2.21 and Fig. 2.22. We assume that the fuzzy peak at 550 nm is associated with the possible presence of a compound such as Alq₃ (HCON(CH)₃₂) in the sample [88].

Based on the measurements of the polarized FL, we calculated the degree of linear polarization ρ using the equation below [137]:

$$\rho = (I_{II} - I_+)/(I_{II} + I)_{,+} \qquad (2.4)$$

where I_{II} and I_+ are the intensities of the perpendicular and parallel luminescence components, respectively. Correction factors from measurements with unpolarized light sources were also used.

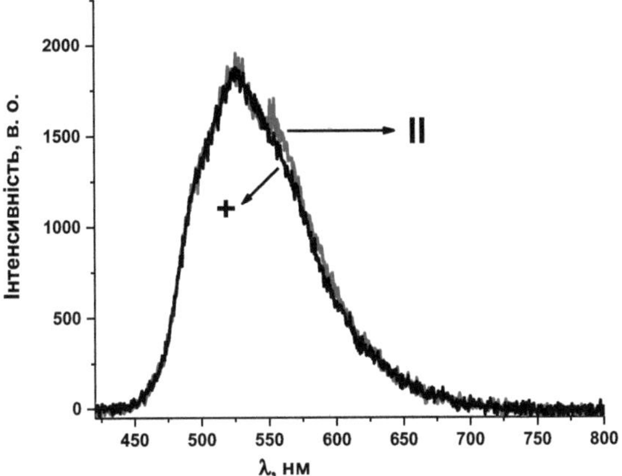

Fig. 2.21. The FTIR spectra of Alq₃ thin film obtained at room temperature, polarized parallel to the X (II) axis and parallel to the Y (+) axis (see Fig. 2.19) (deposition angle 90°) [89].

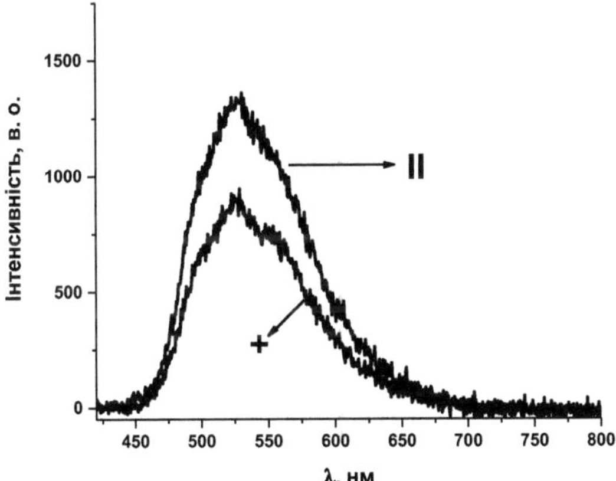

Fig. 2.22. FT-IR spectra of Alq₃ thin film obtained at room temperature, polarized parallel to the X (II) axis and parallel to the Y (+) axis (see Fig. 2.19) (deposition angle 10°) [89].

Using this approach, we found that the degrees of polarization of Alq$_3$ thin films on glass substrates placed at deposition angles of 90° and 10° at a light wavelength of 526 nm were equal to $\rho \approx 0.02$ and $\rho \approx 0.19$, respectively.

As is well known, the Alq$_3$ molecule has a constant dipole moment of 4.1 D [138, 139], and there are many Alq$_3$ molecules per unit volume ($\sim 10^{14}$ cm^{-3}) [139]. The molecules tend to settle in such a way that their dipole moments are directed away from the substrate. Many researchers have observed the presence of a giant surface potential (GSP), which reached +50 V/micron, for Alq$_3$ films deposited in the dark [138-143]. GSP disappears after irradiation with light during or after film deposition if photons are absorbed by Alq$_3$ [140, 142, 144]. The non-centrosymmetric orientation of the molecule is considered to be the cause of GSP [140, 143, 145], but the mechanism of non-centrosymmetry is still unclear. The degree of alignment of molecules is small ($\langle \sin\theta \rangle \sim 0.05$, where θ is the angle between the dipole moment and the layer plane, and $\langle \ \rangle$ is the thermodynamic average value for molecules in all directions) [139]. Therefore, the degree of anisotropy is also low [138, 140]. The small value of the degree of polarization obtained at a deposition angle of 90° in our case is in good agreement with these statements.

In the OAD process, the substrates are positioned at an oblique angle to the vapor flow. During the adsorption of the initial vapor particles on the substrate and the formation of the first islands, the self-shading of these islands does not allow the vapor flow to reach the shaded areas, which leads to an ordered porous and columnar growth [122, 128] (Fig. 2.19 b). In our opinion, this may be the reason for the significant increase in the value of the degree of polarization in our case for the thin film of Alq$_3$ deposited at an angle of 10°.

Thus, by thermal vacuum deposition on glass substrates placed at deposition angles of 90° and 10°, we have obtained Alq_3 thin films. We have presented the optical and morphological properties of these films above, and found that Alq_3 thin films obtained by oblique angle deposition can exhibit significant polarized luminescence.

2.6 Polarized photoluminescence of DCM and its derivatives thin films obtained by oblique angle deposition

Over the past two decades, technologies for the creation of new fluorescent dyes based on 4H-pyrans have been intensively developed, with further study of their properties [115-119, 146-148].

Starting with the pioneering work of Teng and Van Slyke [120, 150] and studies of polymer electroluminescence [151], the study of polarized luminescence in organic structures has quickly established itself as a new field of applied research. The current state of affairs in this relatively new field has been summarized in a number of works, including [1, 2, 129, 152].

In this section of the book, we present the results of our study of the polarized photoluminescence of thin films of dicyanomethylene pyran and its derivatives obtained by thermal vacuum evaporation on glass substrates placed perpendicularly and at an angle of 10° to the vapor flow of the deposited substance. On this topic, we were able to find only two publications [149, 153], the authors of which studied the polarized luminescence of DCM compounds in solvents and in a polymethyl methacrylate (PMMA) matrix. In particular, the authors of [153] found a weak circular polarization of the DCM luminescence.

Thin films of DCM and its derivatives with a thickness of approximately 50 nm on glass substrates placed perpendicularly and at an angle of 10° to the vapor flow of the deposited substance were obtained by thermal vacuum evaporation of powders of starting organic compounds from quartz crucibles at a pressure of 10^{-4} Pa. DCM powder (99.1 % purity) -4-(dicyanomethylene)-2-(4-amylcyclohexyl)-6[4-(dimethylaminoethyl)]-4H-pyran was purchased from Acros Organics (Geel, Belgium). The dicyanomethylene pyran derivatives DCM-5, DCM-17 (C H N_{22233} O) and DCM-18 were synthesized in the laboratory "Materials and Technologies of LCD Devices" and kindly provided for research by the leading researcher Oleksandr Kukhta. The chemical structure of the molecules of the studied thin-film organic compounds is shown in Fig. 2.16 and Fig. 2.23.

The thickness of organic films was monitored online by a quartz film thickness gauge KIT (Akademprylad, Sumy, Ukraine).

The FTIR spectra at room temperature were measured using a quartz polarizer (Glan-Taylor prism) and a portable fiber optic spectrometer from Avantes BV (Apeldoorn, the Netherlands) "AvaSpec-ULS2048L-USB2-UA-RS" with an entrance slit of 25 µm, a diffraction grating of 300 ppm and a resolution of 1.2 nm. The signal accumulation time was 200 ms.

A rotating polarizer was placed between the photoluminescent source and the light detector. Special software was used for automated computer control of the spectrometer and spectrum processing. The samples were excited by the unpolarized light of an M365FP1 fiber-optic-coupled LED (λ = 365 nm, half-width of the diffraction line of maximum (FWHM) - 9 nm, LED output power - 15.5 mW, Thorlabs, Inc.)

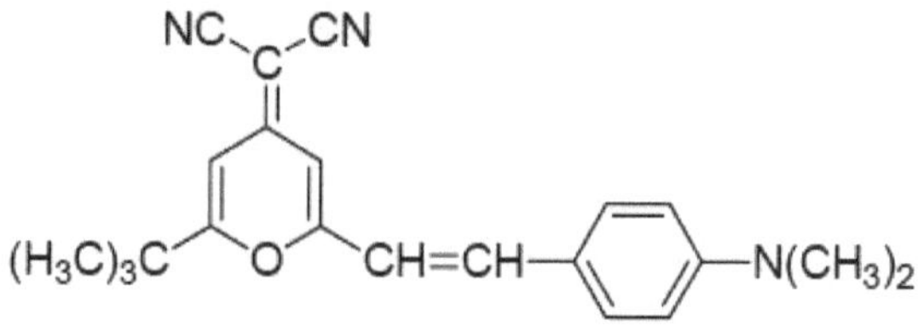

Fig. 2.23. Chemical structure of the DCM-17 molecule [126].

The charge separation in the DCM molecule provides it with a significant dipole moment in both the ground and electronically excited states and determines its electrical properties [125]. For the same reason, the position of the absorption and fluorescence bands depends on the polarity of the solvent [125]. The equilibrium geometry in the ground electronic state helps to understand the possible interaction and arrangement of molecules. Such a geometry for the ground electronic state and the dipole moments of the free DCM molecule were calculated in our previous work using the density functional theory method [154]. The calculated dipole moment of this molecule is $D = 15$ and lies in the plane of the molecule [154].

The polarized PL spectra measured at room temperature of DCM, DCM-5, DCM-17, and DCM-18 thin films are presented in Figs. 2.24-2.27. For all samples, luminescence was observed in the region of light wavelengths 550-800 nm. The obtained spectra are similar to the photoluminescence spectra of DCM films discussed in our previous works [154, 155]. The explanation of the nature of the photoluminescence bands of DCM molecules in various solvents is very well described in [117]. According to [117], the broad photoluminescence band of DCM molecules in solvents can be formed by emission from the locally excited state ("locally excited state", emission wavelength $\lambda = 560\text{-}580$ nm), intramolecular charge transfer emitting state ($\lambda = 610$ nm) and twisted intramolecular charge transfer state ($\lambda = 630$ nm). It should be noted that the luminescence bands of DCM films, compared to the luminescence bands of DCM solutions, are shifted to the red region of the light wavelength spectrum due to a more significant intermolecular interaction [155-157].

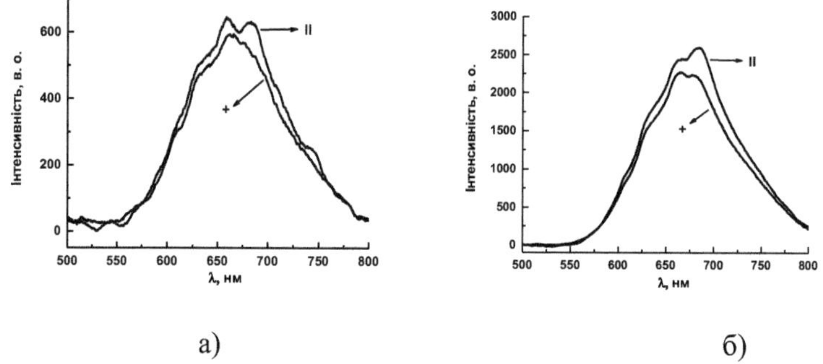

a)   б)

Fig. 2.24. The FTIR spectra measured in the crossed (+) and parallel (||) configurations of the polarizer and the sample at room temperature: a) DCM thin film deposited on a substrate placed perpendicular to the vapor flow, b) DCM thin film deposited on a substrate placed at an angle of 10° to the vapor flow [126].

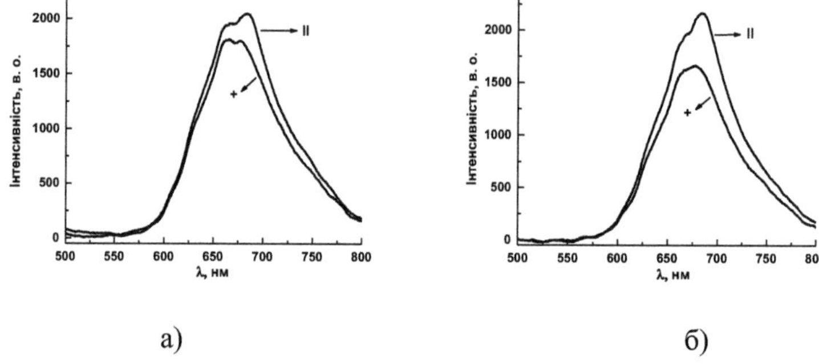

a)   б)

Fig. 2.25. The FTIR spectra measured in the crossed (+) and parallel (||) configurations of the polarizer and the sample at room temperature: a) DCM-5 thin film deposited on a substrate placed perpendicular to the vapor flow, b) DCM-5 thin film deposited on a substrate placed at an angle of 10° to the vapor flow [126].

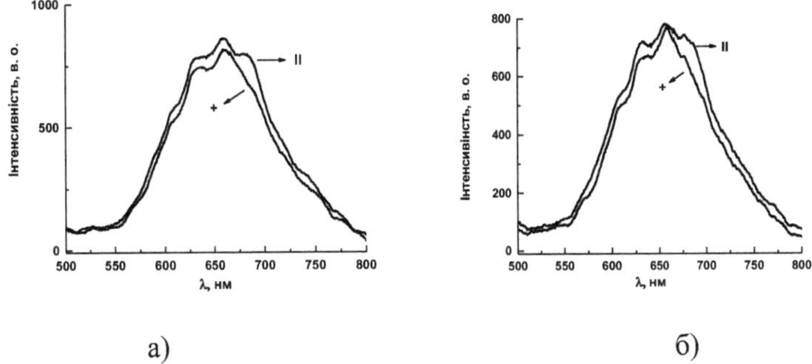

Fig. 2.26. The FTIR spectra measured in the crossed (+) and parallel (||) configurations of the polarizer and the sample at room temperature: a) DCM-17 thin film deposited on a substrate placed perpendicular to the vapor flow, b) DCM-17 thin film deposited on a substrate placed at an angle of 10° to the vapor flow [126].

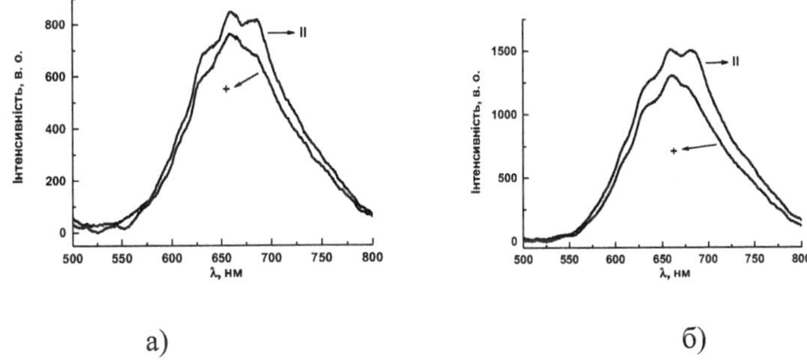

Fig. 2.27. The FTIR spectra measured in the crossed (+) and parallel (||) configurations of the polarizer and the sample at room temperature: a) DCM-18 thin film deposited on a substrate placed perpendicular to the vapor flow, b) DCM-18 thin film deposited on a substrate placed at an angle of 10° to the vapor flow [126].

Based on the measurements of polarized photoluminescence, we calculated the degree of linear polarization ρ using Eq. 2.4 above [137]. The degrees of linear polarization for DCM, DCM-5, DCM-17, and DCM-18 thin films deposited on substrates placed perpendicularly and at an angle of $10°$ to the vapor flow did not exceed the values of, respectively: 0.08 and 0.09; 0.08 and 0.14; 0.06 and 0.05; 0.09 and 0.12. Relatively small values of the degrees of linear polarization and insignificant differences between the degrees of linear polarization of films deposited on substrates placed perpendicularly and at an angle of $10°$ to the vapor flow can be explained by the fact that we are dealing with high molecular weight organic compounds. The lowest values of ρ *were* obtained for DCM-17 films, which may indicate their amorphous structure and the lowest tendency of DCM-17 molecules to aggregate among the studied compounds.

Thus, we have found that the use of the method of deposition of DCM, DCM-5, DCM-17, and DCM-18 films at an oblique angle has a negligible effect on the change in the degree of linear polarization of the luminescence.

2.7. Synthesis and optical properties of composite structures based on ZnO and Alq$_3$

Due to their widespread practical use, OLED technologies are in the center of constant attention of scientists working in the field of optoelectronics. Such serious problems of OLED devices as deterioration of properties due to the influence of oxygen and water molecules, relatively low glass transition temperatures, and low mobility of charge carriers due to the amorphous nature of solids made of organic molecules remain unresolved [4]. In addition, the problem of WOLED manufacturing also remains relevant [158]. One of the promising ways to solve these problems is to use nanocomposites to create hybrid organic-inorganic LEDs. In order to optimize such hybrid systems with controlled optoelectronic properties, it is necessary and important to have a deep understanding of the processes of electronic energy transfer between organic and inorganic subsystems [5].

As already mentioned, Alq3 is widely used in OLED technology. Whereas ZnO-based semiconductor materials are now considered the best alternative to ITO because they are much cheaper and non-toxic [93]. ZnO is also a light-emitting material [159]. There are already many companies in the world that produce transparent and conductive oxides based on ZnO for the needs of electronics [45]. The most challenging problem of ZnO-based electro-optical devices is the lack of stable and reliable *p-type* doping; mainly due to the self-compensation property of ZnO [159]. For this reason, ZnO-based LEDs are made by combining *n-type* ZnO with a *p-type* semiconductor other than ZnO, such as $Cu_2 O$, ZnTe, $SrCu O_{22}$, AlGaN, GaN, or a *p-type* conductive polymer [159, 160]. There are many publications devoted to ZnO and Alq3 separately, while composite materials based on them have been studied by only a few research groups from China, India, Japan, Switzerland, the Republic of Korea, and the United States [90]. They considered different combinations of this composite, such as amorphous or polycrystalline Alq3 or ZnO, doped ZnO or Alq3 , composite nanowires, heterostructures or Alq3 embedded in porous ZnO and observed an enhancement of the photoluminescence intensity of Alq3 and ZnO [90].

In this paragraph, we present data on the fabrication and luminescent properties of composites based on ZnO structures and Alq thin films3 .

A method is known for obtaining a nanocomposite material in powder form based on nanoparticles of zinc oxide and aluminum trichinolinate (Alq3) [161]. According to this method, 4.398 g of zinc acetate dihydrate ($Zn(CH_3 COO)_2$ -$2H_2 O$) is dissolved in 100 ml of ethanol, stirring the solution continuously, and NaOH is added to maintain the pH value at 7. The resulting white ZnO precipitate is washed several times with distilled water. The washed precipitate is dried at 343 K. To obtain the composite Alq3 /ZnO with 10 wt. % ZnO, 4.13 g of Alq3 is dissolved in 100 mL of ethanol and stirred. After two hours of stirring the Alq3 solution, a previously prepared solution of ZnO in ethanol with a concentration of 0.0081 g/liter is added dropwise. The substances are mixed until a homogeneous mixture of both components is formed. The resulting material is dried at 423 K. The disadvantage of the method is the use of ZnO nanoparticles with surfaces covered with OH groups, since their synthesis occurs at a temperature of 423 K. The presence of hydroxyl groups leads to the degradation of the compound Alq_3 and, accordingly, to the rapid destruction of the composite.

Another method of obtaining a nanocomposite film material based on zinc oxide and Alq_3 nanoparticles is known [162]. According to this method, 4.398 g of zinc acetate dihydrate ($Zn(CH_3 COO)_2$ -$2H_2 O$) is dissolved in 100 ml of ethanol, stirring the solution continuously, and NaOH is added to maintain the pH value at 7. The resulting white ZnO precipitate is washed several times with distilled water. The washed precipitate is dried at 343 K. Alq_3 and ZnO (5 wt%, 10 wt%, 20 wt%, 30 wt%, 40 wt% or 50 wt%) are dissolved in 100 ml of ethanol ($C H_{25} OH$) with 0.2 ml of concentrated acid HCl. The solution is stirred continuously at 343 K for an hour. Alq_3 /ZnO films are deposited on a glass substrate. Each of the ten layers of the film is dried in air and annealed at 423 K for five minutes. When a coating of ten 10 layers is obtained, the nanocomposite film material is annealed for an hour at a temperature of 423 K. The disadvantage of the method is the use of ZnO nanoparticles with surfaces coated with OH groups, since their synthesis occurs at a temperature of 423 K. The presence of hydroxyl groups leads to the degradation of the Alq compound$_3$ and, accordingly, to the rapid destruction of the composite.

There is also a known method for preparing a nanocomposite film material based on zinc oxide and Alq_3 [163]. According to this method, the precursor of the ZnO sol-gel is obtained by dissolving 0.5 M zinc acetate dihydrate ($Zn(CH_3 COO)_2$ -$2H_2 O$) in 2-methoxyethanol, with the addition of a stabilizer - monoethanolamine with a ratio of 1:1 and with two-hour stirring at 450 rpm and 333 K. This solution was left at room temperature for 24 hours to form a gel, then filtered through a 0.2 μm syringe filter. 10^{-3} M Alq_3 is dissolved in 2-methoxyethanol, stirring the solution for two hours at 450 rpm. After that, the solution is filtered using a 0.2 μm syringe filter. The filtered solutions of ZnO and Alq_3 are mixed in a volume ratio of 4:1. The resulting substance is deposited by centrifugation on glass and indium tin oxide (ITO) coated substrates at 2000 rpm for 30 s. Five coatings are made. After each coating, the samples are dried at 473 K for 10 min on a hot plate. The disadvantage of the method is the use of ZnO nanoparticles with surfaces coated with OH groups, since their synthesis occurs at a temperature of 473 K. The presence of hydroxyl groups leads to the degradation of the Alq compound$_3$ and, accordingly, to the rapid destruction of the composite.

A method for producing a composite film material based on ZnO and Alq_3 is known [164]. According to this method, a precursor $(C H)_{252}$ Zn and oxygen are dosed into a high-vacuum unit with a plasma reactor in a single stream so that the pressure in the working chamber of the unit changes from $5 \cdot 10^{-5}$ Pa to 0.05 Pa. The Alq_3 molecules are evaporated from a Knudsen-type cell at 513 K onto a silicon or quartz substrate fixed at a distance of 12 cm above the cell at an angle of 10° to the vertical. The deposition rate of the nanocomposite film is controlled by microweighing using a quartz crystal and maintained at 0.4 nm/s. The deposited film is annealed for 15 minutes in a vacuum at 823 K. The disadvantage of the method is the use of complex and expensive equipment to obtain the composite material.

$_3$Another method of obtaining a nanocomposite film material based on ZnO and Alq is known [165]. According to this method, a 75 nm thick ZnO film is deposited on a glass substrate coated with an ITO layer by high-frequency magnetron sputtering of a ZnO target (99.9 %) with a diameter of 152.4 mm and a thickness of 5 mm. Then, a 55 nm thick Alq_3 film is deposited on top of the ZnO film by thermal vacuum evaporation at a pressure of 1.3 mPa without heating the substrate. The disadvantage of this method is the use of complex and expensive equipment to obtain a ZnO film.

Similar to the method used by us is the method of obtaining a nanocomposite material based on ZnO and Alq$_3$ [166]. According to this method, a solution of zinc acetate dihydrate (Zn(CH$_3$ COO)$_2$ -2H$_2$ O) is applied by centrifugation to a substrate coated with an ITO film to form a layer of nuclei for the growth of ZnO nanorods. The nucleation layer is then annealed at 623 K in air for 20 min. The reaction solution for the preparation of ZnO nanorods is prepared by mixing 25 mM zinc nitrate hexahydrate (Zn(NO)$_{32}$ -6H$_2$ O) and 25 mM urotropin (C H N$_{6124}$) in 200 ml of deionized water in a glass vessel. The glass vessel with the solution and the substrate covered with a layer of embryos is placed in a water thermostat heated to 365 K for 15 min. After that, the substrate is removed from the solution, washed in ethanol and deionized water, and air-dried. The Alq$_3$ film is deposited on top of the ZnO nanorods by thermal vacuum evaporation at a rate of 0.03 nm/s. The disadvantage of the method is the use of ZnO nanorods with surfaces coated with OH groups, since their synthesis occurs at a temperature of 365 K. The presence of hydroxyl groups leads to the degradation of the compound Alq$_3$ and the rapid destruction of the composite.

Instead, we obtained ZnO microneedles on Si (100) substrates by the method of gas transport reactions in a horizontal tubular electric furnace in air at a temperature of 500 °C and using pure Zn metal powder (purity > 98 %) [90]. And when a mixture of high-purity powdered zinc oxide, metal zinc, and graphite in a 1:1:1 ratio was used as the starting material for evaporation, hexagonal ZnO microdisks were obtained. To obtain ZnO microdisks, the powder mixture was placed in the sealed end of a quartz tube, while the substrate was placed near the open end. The powder mixture was heated to a temperature of approximately 1050°C, and the substrate was in the temperature range of 700-750°C. After 60 minutes, we turned off the heating of the electric furnace. After spontaneous cooling to room temperature, a homogeneously deposited white layer, i.e., deposited zinc oxide, could be visually observed on the substrate.

We deposited organic layers of Alq_3 with a thickness of less than 50 nm on top of ZnO structures or on a clean silicon substrate in a vacuum at a pressure of 10^{-4} Pa. For evaporation, we used Alq_3 powder with a purity of 99.995 %, purchased from Sigma-Aldrich Corporation. The thickness of the organic films was monitored online by a quartz film thickness gauge KIT (Akademprylad, Sumy, Ukraine).

X-ray diffraction (XRD) measurements were performed by a STOE STADI P diffractometer with a position-sensitive linear detector in the Bragg-Brentano transmission geometry (Cu $K_{\alpha 1}$ radiation with $\lambda = 0.15406$ nm, Ge (111) monochromator, detector scan step: $0.480° \, 2\theta$, accumulation time: 320 s, angular resolution 2θ: 0.015°, 2θ-range: 5°-65°).

The surface morphology and local chemical analysis of the experimental samples were performed using a scanning electron microscope-microanalyzer REMMA-102-02 (SELMI OJSC, Sumy, Ukraine).

The FTIR spectra at room temperature were measured using a portable fiber optic spectrometer (Avantes BV, Apeldoorn, the Netherlands) "AvaSpec-ULS2048L-USB2-UA-RS" with an entrance slit of 25 μm, a diffraction grating of 300 ppm and a resolution of 1.2 nm with or without a quartz polarizer (Glan-Taylor prism). Special software was used for automated computer control of the spectrometer and spectrum processing. The samples were excited either by unpolarized light from an M365FP1 fiber-optic-coupled LED ($\lambda = 365$ nm, half-width of the diffraction line of maximum (FWHM) - 9 nm, LED output power - 15.5 mW, Thorlabs, Inc,

Newton, USA), or by Nd:YAG laser radiation (266 nm, max. output power - 1 µJ, pulse duration < 1 ns, max. repetition rate - 10 kHz).

The X-ray diffractograms of the original Alq₃ powder and the composite material based on Alq₃ and ZnO microdisks are shown in Fig. 2.28. We compared the diffractogram of our favorable powder with the XRD patterns ofα ,β , ,γδ , and ε-Alq₃ [90]. The powder showed a diffractogram quite similar to that of theα -phase of Alq₃ . The peaks obtained for the composite material correspond to the hexagonal pure wurtzite phase of ZnO with lattice parameters a = b = 3.2502± 0.0004 Å, c = 5.2058± 0009 Å [167]. The peak corresponding to the (002) plane was the most intense. No peaks were found from the deposited Alq₃ layer.

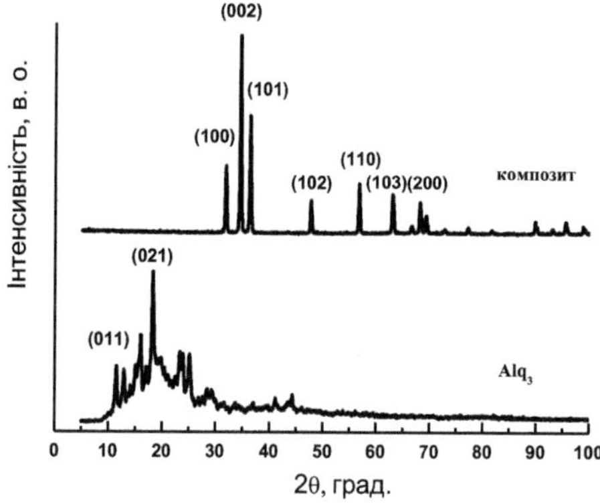

Fig. 2.28. X-ray diffractograms of the original Alq₃ powder (lower curve) and the composite material based on Alq₃ and ZnO microdisks (upper curve)

The morphology of the studied composites is shown in Fig. 2.29. As can be seen, the hexagonal ZnO microdisks with the upper polar surface (0001) were from 2 µm to 8 µm in diameter, and the ZnO microneedles were approximately 1-2 µm in diameter and ~4 µm long.

The FT-IR spectra measured at room temperature of ZnO microneedles (Fig. 2.30) and microdisks (Fig. 2.31) had two bands in the UV and visible regions. The band at 390 nm is typical for ZnO and arises from the recombination of free excitons, bound excitons, and transitions in donor-acceptor pairs [90]. A wide band in the range from about 450 nm to 650 nm is due to defects, primarily uncontrolled impurities and stoichiometry defects [90].

<table>
<tr><td>a)</td><td>б)</td></tr>
</table>

Fig. 2.29. Micrographs of composite structures based on Alq thin films₃ and:
a) ZnO microdisks, b) ZnO microneedles

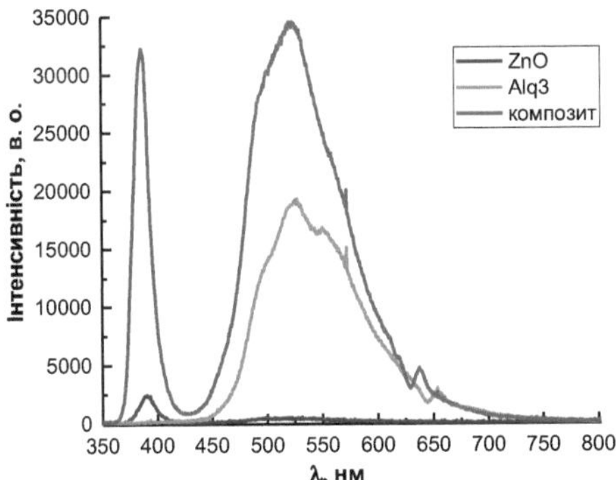

Fig. 2.30. The FTIR spectra measured at room temperature of Alq₃ thin film, ZnO microneedles, and a composite structure based on ZnO microneedles and Alq thin film₃

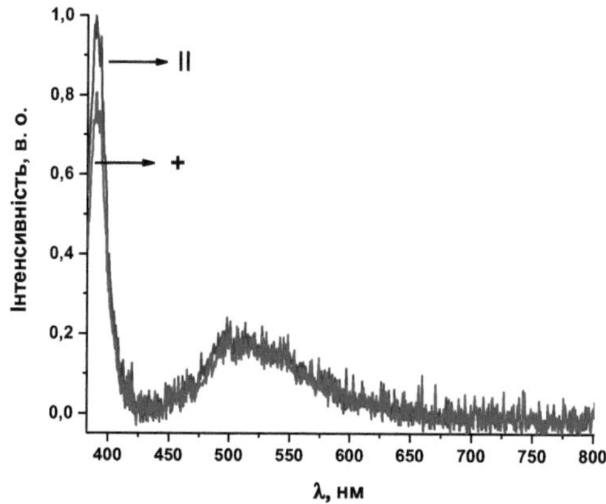

Fig. 2.31. FT-IR spectra of ZnO microdisks measured at room temperature in crossed (+) and parallel (‖) polarizer and sample configurations

The FTIR spectrum of Alq₃ thin film obtained at room temperature showed a characteristic green glow with a maximum at a light wavelength of approximately 525 nm [90] when excited by a laser beam with an emission wavelength of 266 nm (Fig. 2.30). Alq₃ is characterized by crystallization polymorphism both when precipitated from the vapor phase and when evaporated from solution [90]. In the spectrum of the thin film of Alq₃ , a peak at 549 nm was weakly distinguished. We assume that this peak is associated with the possible presence of a compound such as Alq₃ (HCON(CH)₃₂) in the film [9]. In our case, according to [5], the weak peak at approximately 652 nm comes from 8-hydroxyquinoline. It is known that the Alq₃ layer decomposes in the presence of water with the release of 8-hydroxyquinoline [5].

As can be seen from Fig. 2.30, the intensity of the UV-visible FL band of the composite is more than 10 times higher than the intensity of the same band for ZnO microneedles, and the intensity of the band with a maximum at 525 nm is approximately twice as high as the intensity of this band in the FL spectrum of the Alq thin film$_3$. It can be seen that both bands are not very wide and are sufficiently shifted, and the shoulders of both bands do not increase the intensity of each other. These results correlate with the data presented in [160-163]. According to [160-163], the photoluminescence intensity of the composite material based on ZnO and Alq$_3$ has a higher intensity compared to the FL of pure Alq$_3$ and ZnO due to the processes of energy transfer between inorganic and organic materials. According to [163], when a composite material based on ZnO and Alq$_3$ is excited by laser irradiation with a wavelength of 266 nm, the ZnO and Alq$_3$ molecules are excited simultaneously. The excited state energy of the Alq$_3$ molecule can be absorbed by the luminescent quencher, and in turn, the absorbed energy can be transferred without radiation to the ZnO over time, resulting in an increase in the UV emission (edge radiation) of the ZnO in the composite.

Based on the measurements of the polarized FL, we calculated the degree of linear polarization ρ using Eq. (2.4) [137]. It was found that the values of the degrees of polarization in the UV and visible regions of the spectrum for ZnO microdisks are $\rho \approx 0.108$ and $\rho \approx 0$, respectively. Zinc oxide is a semiconductor material with a band gap of 3.37 eV at room temperature [168]. Free excitons in zinc oxide are stable even at room temperature due to their high binding energy (60 meV) [169]. As a result, it is possible to obtain a laser effect based on exciton recombination. Random laser generation with optical pumping has been observed in ZnO micro- and nanostructures [170-172]. Thus, the value of the degree of polarization in the UV region of the spectrum may be due to the presence of the laser generation effect in ZnO microdisks.

Fig. 2.32 shows the FTIR spectrum of a composite structure based on ZnO microdisks and an Alq thin film$_3$ measured at room temperature. The degree of linear polarization of the FL of this composite at a light wave of 521 nm, calculated by eq. (2.4), is $\rho = 0.12$.

In [173], the influence of crystalline and amorphous substrates (KCl (001), KBr (001), ITO, silicon, and glass slide) on the mechanism of formation of Alq$_3$ thin films and their FL anisotropy was studied. The Alq$_3$ films formed on KCl and KBr had needle-like crystals along the <110> directions of the substrate surface. While Alq$_3$ films on ITO and quartz tended to form hemispherical molecular aggregates. Only on crystalline substrates was the photoluminescence anisotropy of Alq$_3$ films observed.

We believe that the Alq$_3$ films retain their intrinsic molecular spacing (0.89 nm) along the <110> direction of the ZnO (001) microdisk surface from the early growth stage. The lattice constants of the ZnO hexagonal cell are $a \approx 3.25$ Å and $c \approx 5.20$ Å [160]. This may be the reason for the higher degree of polarization in the composite material compared to its components.

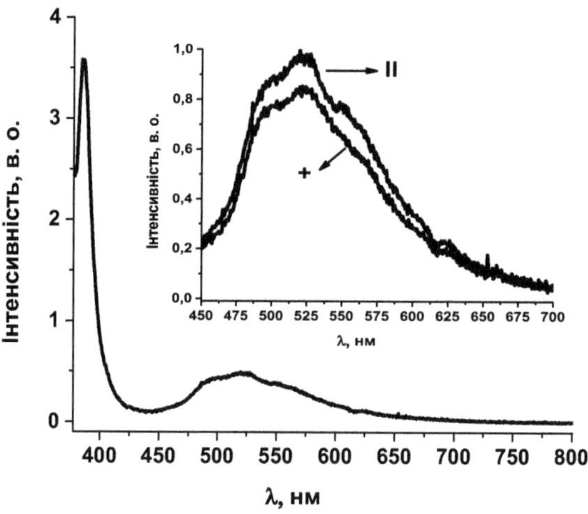

Fig. 2.32. The FT-IR spectrum of the composite based on ZnO microdisks and Alq thin film$_3$ measured at room temperature. Inset - FT-IR

spectra measured at room temperature of the composite in the crossed (+) and parallel (∥) configurations of the polarizer and sample

Thus, the composite material based on Alq_3 and ZnO exhibits stronger green and UV radiation and a higher degree of visible radiation polarization than is typical of the two components of the composite due to the energy transfer processes between inorganic and organic materials. This composite is a possible candidate for the emitting layer of an OLED device.

SECTION 3
PROTOTYPES OF OLED DEVICES BASED ON DCM DERIVATIVES

Achievements in the development of organic light-emitting diodes have so far only partially unlocked the potential of these devices and pointed to new serious scientific and technological challenges. One of the most urgent tasks in the field of OLED technologies today is to produce diode structures with polarized light emission. The development of effective methods for manufacturing such light-generating structures with specified color characteristics, high dichroic coefficient, sufficient energy efficiency and time stability based on organic nanolayers can significantly optimize the design of LEDs. Manufacturing costs can be reduced by eliminating the need for additional polarizing filters, while increasing brightness and contrast, and widening viewing angles.

From a scientific point of view, the main task on the way to solving the described problems is to find the best match between the inorganic nanostructure and the layer of organic luminescent molecules that form the light-generating element. Therefore, our team is engaged in the production and study of organic-inorganic light-emitting structures, which can be used to create innovative OLED devices with polarized radiation for modern optoelectronic systems. The scope of our research includes the coordination of starting organic and inorganic compounds in multilayer configurations with different hierarchies to create new organic-inorganic structures with polarized luminescence.

An important aspect of our work is the use of newly synthesized highly luminescent organic molecules in the light-generating layers of OLED structures.

The ability to combine specific substrate materials with different organic molecules to emit light at different wavelengths can be seen as a way to create hybrid nanostructures that emit white light.

From the research point of view, this section presents data on the mechanisms of formation of hybrid nanostructures and electronic processes in such structures, which lay the foundation for energy-efficient and controlled methods of creating promising functional materials for optoelectronics, as well as for predicting the impact of external factors on the characteristics of OLED structures.

3.1. Electroluminescence of the heterostructure with ITO/Alq configuration$_3$:DCM-5 (10 wt. %)/Alq /Al$_3$

We thermally deposited DCM-5, Alq$_3$ and Alq$_3$:DCM-5 organic films (10 wt. %) with a thickness of less than 50 nm in a vacuum at a pressure of 10^{-4} Pa on glass substrates that are optically transparent for light wavelengths greater than 300 nm or on glass substrates coated with ITO film (purchased from Sigma-Aldrich, surface resistance - 70-100 Ohm/sq) for OLED fabrication. The thickness of the organic films was monitored online using a quartz film thickness gauge KIT (Akademprylad, Sumy, Ukraine). In the manufacture of OLEDs, the contact as the top electrode - cathode to the Alq$_3$ film was formed by thermal vacuum evaporation of aluminum [174] in a vacuum universal post VUP-5M (SELMI OJSC, Sumy, Ukraine). Aluminum contacts in the form of circles with a diameter of approximately 2 mm were deposited using mask technology. A liquid photoresist-insulator based on o-naphthoquinone diazide and novolac (a type of phenol-formaldehyde resin) "Positive 20" by KONTAKT CHEMIE was used as a photoresist-insulator. A schematic representation of OLED is shown in Fig. 3.1.

Ex situ ellipsometry measurements were performed with a serial ellipsometer LEF-3 M in the polarizer-compensator-sample-analyzer mode. The light source was a He-Ne laser (λ = 632.8 nm).

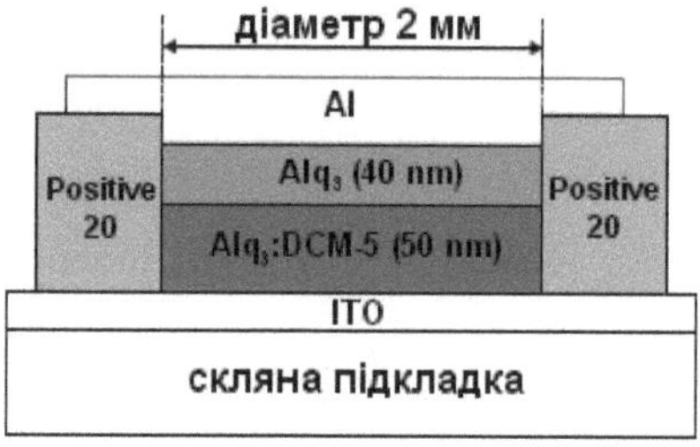

Fig. 3.1. Schematic structure of OLEDs based on Alq₃ and Alq₃ :DCM-5 thin films (10 wt. %)

The FT-IR spectrum of DCM-5 film was measured by an automated monochromator/spectrograph M266 (SolarLS JSC) connected to a CCD camera based on a Hamamatsu S7030-1006S sensor. The sample was excited by a GaN laser (405 nm). The sample was placed in a closed-circuit helium cryostat equipped with a DE-202A cryocooler (Advanced Research Systems, Macungie, USA) and a Cryocon 32 thermostat (Cryogenic Control Systems Inc., Rancho Santa Fe, USA).

The electroluminescence spectra were measured by a portable fiber optic spectrometer from Avantes BV (Apeldoorn, the Netherlands) "AvaSpec-ULS2048L-USB2-UA-RS" with an entrance slit of 200 μm, a diffraction grating of 300 ppm and a resolution of 9 nm. The signal accumulation time was 200 ms. Light detection in the spectrometer is performed by a 2048-pixel CCD matrix. Special software was used for automated computer control of the spectrometer and processing of the spectra.

Ex situ ellipsometric measurements were performed using a single-layer model of DCM-5 thin films on optical glass substrates, and it was found that the effective refractive index of DCM-5 thin films is approximately 2.26.

We used low-temperature studies because the FL of thin films of pure DSM or its derivatives is very weak [120] due to the large intermolecular interaction that quenches luminescence [121, 149]. As can be seen from Fig. 3.2, the FTIR spectrum of DCM-5 thin film measured at T = 96 K contained two overlapping bands with maxima at 635 nm and 665 nm in the visible region.

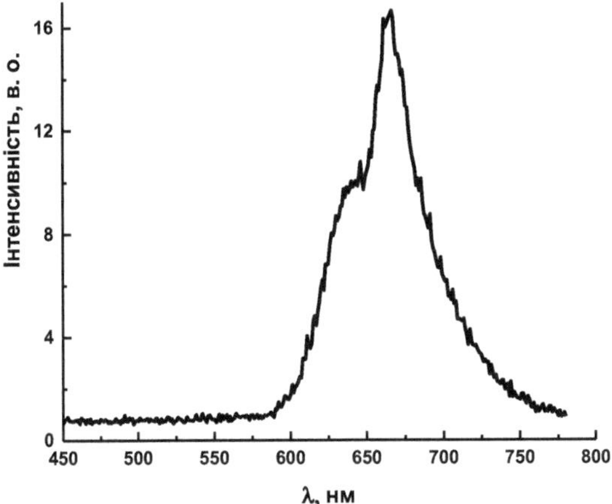

Fig. 3.2. The FTIR spectrum of DCM-5 thin film measured at T = 96 K

It was found that the decay kinetics of these two bands at room temperature consists of two components - fast and slow (Fig. 3.3, Table 3.1).

Table 3.1

Parameters of the attenuation curves of the FL bands of DCM-5 thin film under excitation by a GaN laser (405 nm) measured at room temperature

Luminescence band, nm	Fast component decay time, τ_1 , ns	Amplitude of the fast component, A_1	Decay time of the slow component, τ_2 , ns	Amplitude of the slow component, A_2
635	3,2	543	7.4	44
665	3	571	7	40

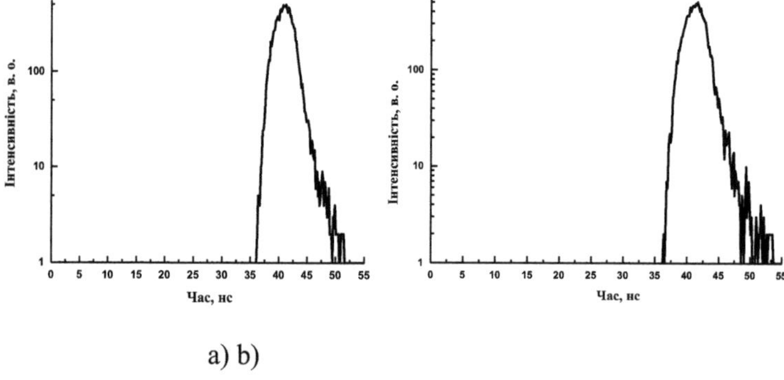

a) b)

Fig. 3.3. Decay time curves of the FL bands of DCM-5 thin film under excitation by GaN laser (405 nm) measured at room temperature: a) bands with a maximum at 635 nm; b) bands with a maximum at 665 nm

The parameters of the FL decay time curves of the DCM-5 thin film are consistent with the data given in [175].

The problem with the weak FL of the DCM-5 thin film was overcome by using the "guest-host" system. That is, by mixing molecules of the "guest" compound with another substance - the "host". We used the organic substance Alq_3 , which is usually used as a host compound due to the compatibility of energy levels [176-178].

Fig. 3.4 shows the electroluminescence spectra of our heterostructure with the configuration ITO/Alq$_3$:DCM-5 (10 wt. %)/Alq$_3$ /Al. The electroluminescent emission was clearly visible to the naked eye in the dark (see inset in Fig. 3.4). The intensity of the electroluminescence increased with the applied voltage. The broad band in the range from about 600 nm to 750 nm has a clear shoulder, at about 530 nm, on the high energy side. According to the literature, under normal conditions, Alq$_3$ emits green light with a maximum at about 530 nm [179].

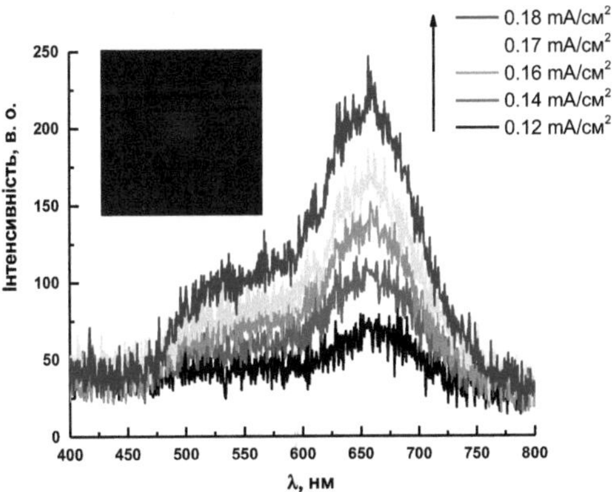

Fig. 3.4. Electroluminescence spectra of OLEDs with ITO/Alq configuration$_3$:DCM-5 (10 wt. %)/Alq$_3$ /Al measured at room temperature at different injection currents. Inset - photo of luminescence

The electroluminescent radiation of the ITO/Alq heterostructure$_3$:DCM-5 (10 wt. %)/Alq$_3$ /Al (for a current density of 0.18 mA/cm^2) had CIE (x; y) color coordinates (0.43; 0.40) (Fig. 3.5). That is, warm white light with an equivalent temperature of T = 3050 K was emitted.

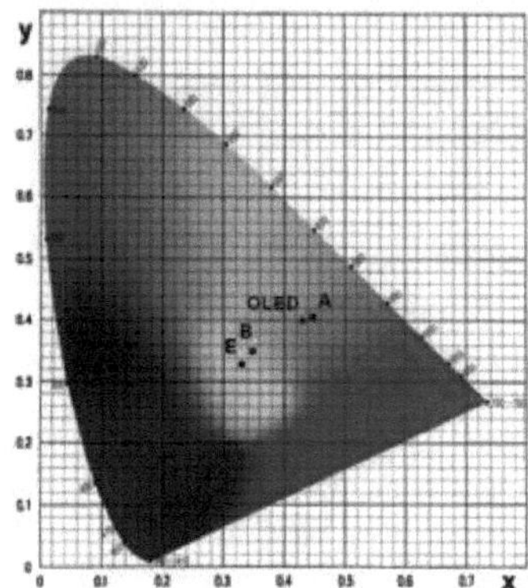

Fig. 3.5. The electroluminescence spectrum of the ITO/Alq$_3$:DCM-5 (10 wt. %)/Alq$_3$ /Al heterostructure in color coordinates: OLED - our heterostructure; A, B, E - standard emitters

3.2. Electroluminescence of the heterostructure with ITO/Alq configuration$_3$:DCM-18 (10 wt. %)/Alq /Al$_3$

In this paragraph, we present the results of optical spectral studies of thin films and OLED structures obtained by thermal vacuum evaporation of Alq$_3$ powders and the dicyanomethylene pyran derivative DCM-18 (C H N$_{252133}$) O. Since we were unable to find publications on OLED devices based on DCM-18.

The methods for obtaining experimental samples were the same as in Section 3.1.

The spectra of photo- and electroluminescence at room temperature were measured by a portable fiber optic spectrometer from Avantes BV (Apeldoorn, the Netherlands) "AvaSpec-ULS2048L-USB2-UA-RS" with an entrance slit of 200 µm, a diffraction grating of 300 ppm and a resolution of 9 nm. The signal accumulation time was 200 ms. Special software was used for automated computer control of the spectrometer and spectrum processing. To obtain the FTIR of DCM-18 thin film, light from an

M365FP1 fiber-optic-coupled LED ($\lambda = 365$ nm, FWHM - 9 nm, LED output power - 15.5 mW, Thorlabs, Inc., Newton, USA) was used as an excitation source.

Fig. 3.6 shows the FTIR spectrum of DCM-18 thin film measured at room temperature. In fact, the FTIR spectrum is a single broad band in the range from about 540 nm to 800 nm, with a clear shoulder (with a maximum at about 575 nm) on the high energy side.

Fig. 3.7 shows the electroluminescence spectra of our heterostructure with the configuration ITO/Alq$_3$:DCM-18 (10 wt. %)/Alq$_3$ /Al. The intensity of electroluminescence increased with increasing applied voltage. The electroluminescence spectra contained a single broad band in the range from about 550 nm to 800 nm with a maximum in the vicinity of 660-665 nm.

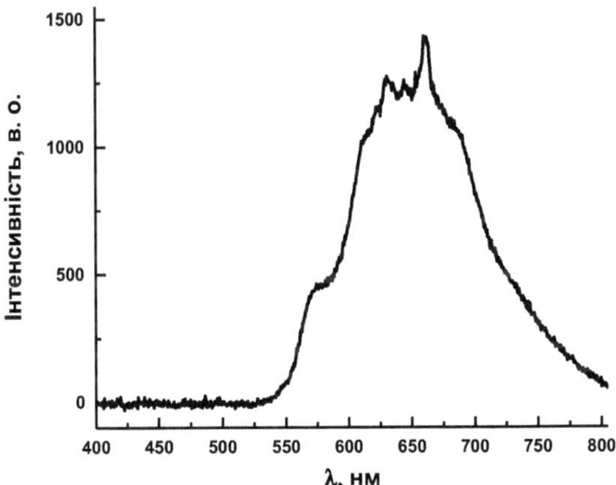

Fig. 3.6. FTIR spectrum of DCM-18 thin film obtained at room temperature

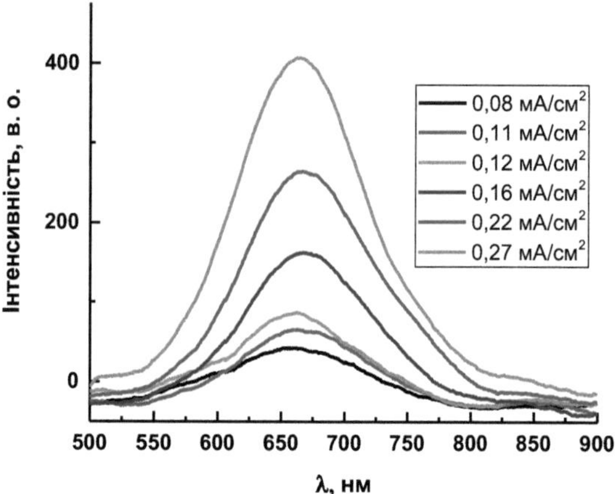

Fig. 3.7. Electroluminescence spectra of OLEDs with ITO/Alq configuration3 :DCM-18 (10 wt. %)/Alq3 /Al measured at room temperature at different injection currents

The electroluminescent radiation of the ITO/Alq heterostructure3 :DCM-18 (10 wt. %)/Alq3 /Al (for a current density of 0.27 mA/cm^2) had CIE (x; y) color coordinates (0.56; 0.32) (Fig. 3.8). That is, hot red-orange light with an equivalent temperature of T = 2112 K was emitted.

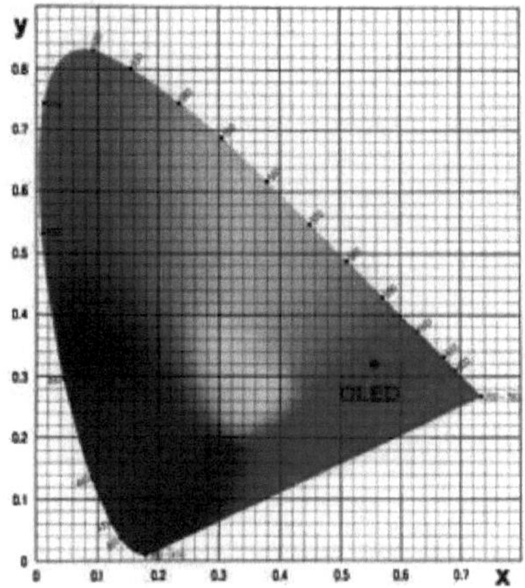

Fig. 3.8. Electroluminescence spectrum of ITO/Alq$_3$:DCM-18 (10 wt. %)/Alq$_3$ /Al OLED heterostructure in color coordinates

CONCLUDING REMARKS, CONCLUSIONS AND PERSPECTIVES

Here is a brief list of the most interesting original scientific results presented in the monograph and the conclusions that can be drawn from the analysis of the large array of data under consideration.

It was found that low-temperature annealing for one hour and short-term irradiation with light from a quartz mercury lamp lead to an increase in the average size of crystallites and the root mean square surface roughness of ITO films. The lowest optical transparency in the visible range of the spectrum among the studied experimental samples was observed in the unannealed and non-irradiated ITO films, and the highest in the annealed at the highest temperature (150°C) and irradiated for 5 min with light from a DRT-125 lamp. An increase in the irradiation time leads to a decrease in the optical transparency of the films, which may be due to an increase in the number of defects due to the processes of ITO photodecomposition and ablation. Thermal annealing and irradiation with light from a quartz mercury lamp reduce the reflection coefficient of ITO films in the visible spectral range. Based on the analysis of the optical absorption spectra of ITO films, an increase in the value of their optical band gap with an increase in the annealing temperature was found. On the other hand, the dependence of the optical bandgap of ITO films on the time of irradiation with light from a quartz mercury lamp turned out to be non-trivial, which is associated with changes in the concentration of structural defects. With an increase in the annealing temperature and with an increase in the time of irradiation with a quartz mercury lamp, a decrease in the value of the resistivity of ITO films was recorded.

The absorption properties of Alq_3 thin film in the UV and visible wavelengths of the light spectrum were studied for the first time in the temperature range of 16-320 K.

By depositing DCM, DCM-5, DCM-18, Alq_3 :DCM (10 wt% DCM), and Alq_3 :DCM-5 (10 wt% DCM-5) films at an oblique angle, the refractive indices can be manipulated.

Alq_3 thin films obtained by oblique angle deposition can exhibit significant polarized luminescence.

It was found that the use of the method of deposition of DCM, DCM-5, DCM-17, and DCM-18 films at an oblique angle has a negligible effect on the change in the degree of linear polarization of the luminescence.

The composite material based on Alq_3 and ZnO exhibits stronger green and UV radiation and a higher degree of visible radiation polarization than is typical of the two components of the composite due to the energy transfer processes between inorganic and organic materials. This composite is a possible candidate for the emitting layer of an OLED device.

We have also fabricated and studied the optical properties of organic LEDs with the configurations ITO/Alq_3 :DCM-5 (10 wt%)/Alq_3 /Al and ITO/Alq_3 :DCM-18 (10 wt%)/Alq_3 /Al.

These results will be useful for the optimization and design of highly efficient organic LEDs.

The ever-increasing demand for high-performance display technologies in consumer electronics is driving the development and synthesis of a variety of new OLED materials. Although scientists and engineers have accumulated considerable knowledge about the laboratory synthesis of these materials in recent years, the process of experimental fabrication and study of synthesized structures still requires a significant amount of effort and time [180].

Atomic-scale modeling has been and is an important tool for navigating the vast chemical space of new OLED materials. With the increasing computing power of computers, massive theoretical screening of millions of compounds has become a reality [180]. Continuous improvement and development of neural network modeling methods and automated machine learning algorithms will further significantly reduce the cost of developing new organic materials.

Looking to the future, structure simplification, design flexibility, and cost reduction will remain the main factors for further large-scale industrialization of OLEDs. Therefore, many scientists believe that ultra-thin (less than 1 nm) light-emitting layers of phosphorescent dyes will be more widely used in OLED devices [181]. Technologies for full 3D printing of OLEDs are also being developed [182].

The authors hope that the problems raised in the monograph concerning the synthesis and study of low-dimensional organic and inorganic materials will be useful in the creation of new micro- and nanoelectronic devices.

BIBLIOGRAPHY

1. Zhou L. Highly Linearly Polarized Light Emission from Flexible Organic Light-Emitting Devices Capitalized on Integrated Ultrathin Metal-Dielectric Nanograting / L. Zhou, Y.-F. Zhu, Q.-Y. Zhang [et al.] // Optics Express. - 2020. - V. 28. - P. 13826-13836.

2. Organic Light Emitting Devices / ed. by J. Singh. - Rijeka: TechOpen, 2012. - 244 p.

3. Dyreklev P. Polarized Electroluminescence from an Oriented Substituted Polythiophene in a Light Emitting Diode / P. Dyreklev, M. Berggren, O. Inganas [et al.] 7. - P. 43-45.

4. Organic Light-Emitting Diodes (OLEDs): Materials, Devices and Applications / Ed. by A. Buckley. - Sawston: Woodhead Publishing Limited, 2013. - 666 p.

5. Manzhia P. Mg-Doped ZnO Nanostructures for Efficient Organic Light Emitting Diode / P. Manzhia, R. Kumari, M. B. Alam [et al. // Vacuum. - 2019. - V. 166. - P. 370-376.

6. Chen Y. Highly Efficient Inverted Organic Light-Emitting Devices Adopting Solutionprocessed Double Electron-Injection Layers / Y. Chen, S. Chu, R. Li [et al.] - 2019. - V. 66. - P. 1-6.

7. Lee H. Improvement of Electron Injection in Inverted Bottom-Emission Blue Phosphorescent Organic Light Emitting Diodes Using Zinc Oxide Nanoparticles / H. Lee, I. Park, J. Kwak [et al.] Phys. Lett. - 2010. - V. 96. - P. 153306.

8. Mucur S. P. Triangular-Shaped Zinc Oxide Nanoparticles Enhance the Device Performances of Inverted OLEDs / S. P. Mucur, T. A. Tumay, S. Birdoğan [et al.] 1. - P. 7-14.

9. Qian L. Electroluminescence from Light-Emitting Polymer/ZnO Nanoparticle Heterojunctions at Sub-Bandgap Voltages / L. Qian, Y.

Zheng, K. R. Choudhury [et al. // Nano Today. - 2010. - V. 5. - P. 384-389.

10. Jeong K. W. Enhancing the Electroluminescence of OLEDs by Using ZnO Nanoparticle Electron Transport Layers That Exhibit the Auger Electron Effect / K. W. Jeong, H. S. Kim, G. R. Yi, C. K. Kim // Molecular Crystals and Liquid Crystals. - 2018. - V. 663. - P. 61-70.

11. Kovalev D. S. Dielectric Properties of Liquid-Crystal Azomethine Polymer with a Side Alkyl-Substituted Chain, Doped with Fullerene C60 / D. S. Kovalev, S. V. Kostromin, V. Musteaţa [et al.] Solid State. - 2016. - V. 58. - P. 835-839.

12. Junaid M. A Review on Graphene-Based Light Emitting Functional Devices / M. Junaid, M. H. Md Khir, G. Witjaksono [et al.] 25. - P. 4217.

13. Hong G. A Brief History of OLEDs - Emitter Development and Industry Milestones / G. Hong, X. Gan, C. Leonhardt [et al.] 2021. - P. 2005630.

14. Salehi A. Recent Advances in OLED Optical Design / A. Salehi, X. Fu, D.-H. Shin, F. So // Adv. Funct. Mater. - 2019 - V. 29. - P. 1808803.

15. Kumar S. A. Device Engineering Aspects of Organic Light-Emitting Diodes (OLEDs) / S. A. Kumar, J. S. Shankar, B. K Periyasamy, S. K. Nayak // Polymer-Plastics Technology and Materials. - 2019. - V. 58. - P. 1597-1624.

16. Das H. S. Organic Light Emitting Diode (OLED) Technology and its Applications: Literature Review / H. S. Das, P. K. Nandi // Biodiversity Online Journal. - 2021. - V. 2. - P. 1-3.

17. Ha J. M. Recent Advances in Organic Luminescent Materials with Narrowband Emission / J. M. Ha, S. H. Hur, A. Pathak // NPG Asia Materials. - 2021. - V. 13. - P. 53.

18. Schols S. Device Architecture and Materials for Organic Light-Emitting Devices / S. Schols. - Dordrecht: Springer, 2011. - 154 p.

19. Kalyani N. T. Principles and Applications of Organic Light Emitting Diodes (OLEDs) / N. T. Kalyani, H. Swart, S. J. Dhoble: Woodhead Publishing, 2017. - 319 p.

20. Tsujimura T. OLED Display Fundamentals and Applications. Second Edition / T. Tsujimura. - Hoboken: John Wiley & Sons Inc., 2017. - 320 p.

21. https://omdia.tech.informa.com/OM022509/OLED-Display-Market-Tracker--3Q21-Analysis

22. Baryshnikov G. V. Elements and devices of organic electronics: a collective monograph / G. V. Baryshnikov, I. I. Gelzhynsky, Z. Y. Gotra, H. B. Ivanyuk, B. P. Minaev, P. Y. Stakhira - Lviv: "Prostir-M, 2020. 224 p.

23. Organic light-emitting structures: a collective monograph / G. V. Baryshnikov, I. I. Gelzhynskyi, Z. Y. Gotra, H. B. Ivaniuk, B. P. Minaev, P. Y. Stakhira - Lviv: Lviv Polytechnic Publishing House, 2020. - 236 p.

24. Organic electronics: a textbook / G.V. Baryshnikov, D.Y. Volyniuk, I.I. Gelzhynskyi [et al. Lviv : Lviv Polytechnic Publishing House, 2014. 292 p.

25. Korotun A.V. Nanophotonic technologies. Current state and prospects / A.V. Korotun, A.O. Koval, A.A. Krykin [et al: FOP Sabov A. M., 2019. 482 p.

26. Koval VM Optoelectronic information systems: Lecture notes [Electronic resource]: a textbook for students majoring in 153 "Micro- and nanosystems engineering", educational program "Micro- and nanoelectronics" / V. M. Koval - Kyiv : Igor Sikorsky Kyiv Polytechnic Institute, 2020. 165 p.

27. Vodzinskiy V.Y. Organic semiconductors. Textbook / V. Y. Vodzinsky. - Nizhny Novgorod : Nizhny Novgorod State Technical University named after R. E. Alekseev, 2015. - 72 p.

28.https://electronics.howstuffworks.com/oled3.htm

29.D. Kaushika P. A New Generation of Organic Light-Emitting Displays-OLED / P. Kaushika D., N. Patel and Ravi Vataliya // International Journal of Scientific Research and Reviews. - 2018. - V. 7. - P. 697-706.

30.Development of organic nanoscale light-emitting structures of white color [Text]: dissertation for the degree of Doctor of Technical Sciences in specialty 05.27.01 "Solid-state electronics" / I. I. Gelzhynskyi - Lviv Polytechnic National University, Ministry of Education and Science of Ukraine, Lviv, 2021. 295 p.

31.Fecková M. Organic and Organometallic Heterocyclic Luminescent Materials: Towards OLED Applications. Dissertation Work / M. Fecková. - Université Rennes 1; Univerzita Pardubice, 2021. - 185 p.

32.ZnO as Multifunctional Material for Nanoelectronics. 2nd Supplemented Edition / Turko B., Kapustianyk V. - Beau Bassin: Scholars' Press, 2020. - 160 p.

33.Shigesato Y. ITO thin-film transparent conductors: Microstructure and processing / Y. Shigesato, I. Yasui, D. C. Paine // JOM: the journal of the Minerals, Metals & Materials Society. - 1995. - V. 47. - P. 47-50.

34.Sofi A. H. Structural, Optical and Electrical Properties of ITO Thin Films / A. H. Sofi, M. A. Shah, K. Asokan // Journal of Electronic Materials. - 2018. - V. 47. - P. 1344-1352.

35.Kim J. High-Temperature Optical Properties of Indium Tin Oxide Thin-Films / J. Kim, S. Shrestha, M. Souri [et al.] 10. - P. 12486 (8 pp).

36.Hacini A. Optimization of ITO Thin Film Properties as a Function of Deposition Time Using the Swanepoel Method / A. Hacini, A. H. Ali, N. N. Adnan // Optical Materials. - 2021. - V. 120.- P. 111411 (10 pp).

37.Sibin K. P. Optical and Electrical Properties of ITO Thin Films Sputtered on Flexible FEP Substrateas Passive Thermal Control System for Space

Applications / K. P. Sibin, N. Swain, P. Chowdhury [et al.] 145. - P. 314-322.

38. Rozati M. Transparent Conductive Sn-Doped Indium Oxide Thin Film Deposited by Spray Pyrolysis Techniques / M. Rozati, T. Ganj // Renew. Energy. - 2004. - V. 29. - P. 1671-1676.

39. Oladigbo O. E. Review on Transparent Conductive Oxides Thin Films Deposited by Sol-Gel Spin Coating Technique / O. E. Oladigbo, O. Adedokun, Y. K. Sanusi // International Journal of Engineering Science and Application. - 2018. -V. 2. - P. 88-97.

40. Joshi S. M. Effect of Annealing Atmosphere (Ar vs. Air) and Temperature on the Electrical and Optical Properties of Spin-Coated Colloidal Indium Tin Oxide Films / S. M. Joshi, R. A. Gerhardt // Journal of Materials Science. - 2013. 48. - P. 1465-1473.

41. Jeong J. A. Ink-Jet Printed Transparent Electrode Using Nano-Size Indium Tin Oxide Particles for Organic Photovoltaics / J. A. Jeong, J. Lee, H. Kim [et al.] 94. - P. 1840-1844.

42. Tahar R. B. H. Tin Doped Indium Oxide Thin Films: Electrical Properties / R. B. H. Tahar, Y. Ohya, T. Ban, Y. Takahashi // J. Appl. Phys. - 1998. - V. 83. - P. 2631-2645.

43. Chen Z. High Mobility Indium Tin Oxide Thin Film and Its Application at Infrared Wavelengths: Model and Experiment / Z. Chen, Y. Zhuo, W. Tu [et al.] 26. - P. 22123-22134.

44. Panasyuk M. R. Manufacturing Technology, Optical and Spectral Properties of Nano-Structured Thin ZnO Films / M. R. Panasyuk, B. I. Turko, V. B. Kapustianyk [et al. // Functional Mater. - 2005. - V. 12. - P. 746-749.

45. Turko B. Effect of Dopant Concentration and Crystalline Structure on the Absorption Edge in ZnO:Y Films / B. Turko, U. Mostovoy, M. Kovalenko [et al.] Opt. - 2021. - V. 22. - P. 31-37.

46. Lee K.-S. A Study on Transparent Electrode Properties of Indium Tin Oxide Thin Films Deposited from Recycled Target / K.-S. Lee, Y. J. Mo, I.-K. Park [et al.] Korean Chem. Soc. - 2020. - V. 41. - P. 341-347.

47. Sputtered Indium Tin Oxide Films for Optoelectronic Applications. Chapter 14 in Book: Optoelectronics - Advanced Device Structures / O. Malik and F. J. de la Hidalga-Wade. - Rijeka: IntechOpen CY, 2017. - 372 p.

48. Fikry M. Superior Control for Physical Properties of Sputter Deposited ITO Thin-Films Proper for Some Transparent Solar Applications / M. Fikry, M. Mohie, M. Gamal [et al.] - 2021. - V. 53. - P. 122 (16 pp).

49. Khusayfan N. M. Study of Structure and Electro-Optical Characteristics of Indium Tin Oxide Thin Films / N. M. Khusayfan, M. M. El-Nahass // Advances in Condensed Matter Physics. - 2013. - V. 2013. - P. 408182 (8 pp).

50. Txintxurreta J. Indium Tin Oxide Thin Film Deposition by Magnetron Sputtering at Room Temperature for the Manufacturing of Efficient Transparent Heaters / J. Txintxurreta, E. G-Berasategui, R. Ortiz [et al.] 11. - P. 92 (14 pp).

51. Al-Dahoudi N. Wet Coating Deposition of ITO Coatings on Plastic Substrates / N. Al-Dahoudi, M. A. Aegerter // Journal of Sol-Gel Science and Technology. - 2003. - V. 26. - P. 693-697.

52. Wiboonsak S. Simple ITO Surface Treatments Induced Better Performance for Low Cost Organic Solar Cells / S. Wiboonsak, K. Lohawet, B. A. T. Duong [et al.] 45. - P. 2178-2189.

53. Santos E. R. Low Cost UV-Ozone Reactor Mounted for Treatment of Electrode Anodes Used in P-OLEDs Devices / E. R. Santos, J. I. Balbino de Moraes, C. M. Takahashi [et al.] 26. - P. 236-241.

54. So S. K. Surface Preparation and Characterization of Indium Tin Oxide Substrates for Organic Electroluminescent Devices / S. K. So, W. K. Choi, C.H. Cheng [et al.] Phys. A. - 1999. - V. 68. - P. 447-450.

55. Kim B.-S. UV-Ozone Surface Treatment of Indium-Tin-Oxide in Organic Light Emitting Diodes / B.-S. Kim, D.-E. Kim, Y.-K. Jang [et al. // Journal of the Korean Physical Society. - 2007. - V. 50. - P. 1858-1861.

56. Kim S. Y. Effect of Ultraviolet-Ozone Treatment of Indium-Tin-Oxide on Electrical Properties of Organic Light Emitting Diodes / S. Y. Kim, J.-L. Lee, K.-B. Kim, Y.-H. Tak // J. Appl. Phys. - 2004. - V. 95. - P. 2560-2563.

57. Destruel P. Influence of Indium Tin Oxide Treatment Using UV-Ozone and Argon Plasma on the Photovoltaic Parameters of Devices Based on Organic Discotic Materials / P. Destruel, H. Bock, I. Seguy [et al.] Int. - 2006. - V. 55. - P. 601-607.

58. Aleksandrova M. P. Influence of UV Treatment on the Electrooptical Properties of Indium Tin Oxide Films Used in Flexible Displays / M. P. Aleksandrova, I. N. Cholakova, G. K. Bodurov [et al. // World Academy of Science, Engineering and Technology. - 2012. - V. 71. - P. 1271-1274.

59. Bhatti M. T. Effect of Annealing on Electrical Properties of Indium Tin Oxide (ITO) Thin Films / M. T. Bhatti, A. M. Rana, A. F. Khan, M. I. Ansari // Pakistan Journal of Applied Sciences. - 2002. - V. 2. - P. 570-573.

60. Plumley J. B. Crystallization of Electrically Conductive Visibly Transparent ITO Thin Films by Wavelength-Range-Specific Pulsed Xe Arc Lamp Annealing / J. B. Plumley, A. W. Cook, C. A. Larsen [et al. // Journal of Materials Science. - 2018. - V. 53. - P. 12949-12960.

61. Toporovska L. Zinc Oxide: Reduced Graphene Oxide Nanocomposite Film for Heterogeneous Photocatalysis / L. Toporovska, B. Turko, M. Savchak [et al.] - 2020. - V. 52. - P. (12 pp).

62.Rietveld G. DC Conductivity Measurements in the Van Der Pauw Geometry / G. Rietveld, Ch. V. Koijmans, L. C. A. Henderson [et al. // IEEE Transactions on Instrumentation and Measurement. - 2003. - V. 52. - P. 449-453.

63.Guillen C. Influence of Oxygen in the Deposition and Annealing Atmosphere on the Characteristics of ITO Thin Films Prepared by Sputtering at Room Temperature / C. Guillen, J. Herrero // Vacuum. - 2006. - V. 80. - P. 615-620.

64.Her S.-C. Fabrication and Characterization of Indium Tin Oxide Films / S.-C. Her, C.-F. Chang // J. Appl. Biomater. Funct. Mater. - 2017. - V. 15. - P. e170-e175.

65.Ahmed N. M. The Effect of Post Annealing Temperature on Grain Size of Indium-Tin-Oxide for Optical and Electrical Properties Improvement / N. M. Ahmed, F. A. Sabah, H. I. Abdulgafour [et al.] 13. - P. 102159 (6 pp).

66.Kim H. Indium Tin Oxide Thin Films for Organic Light-Emitting Devices / H. Kim, A. Pique, J. S. Horwitz [et al.] Phys. Lett. - 1999. - V. 74. - P. 3444-3446.

67.Malliaras G. G. The Roles of Injection and Mobility in Organic Light Emitting Diodes / G. G. Malliaras, J. C. Scott // J. Appl. Phys. - 1998. - V. 83. - P. 5399.

68.de Jong M. P. Stability of the Interface Between Indium-Tin-Oxide and Poly(3,4-ethylenedioxythiophene)/Poly(styrenesulfonate) in Polymer Light-Emitting Diodes / M. P. de Jong, L. J. van IJzendoorn and M. J. A. de Voigt // Appl. Phys. Lett. - 2000. - V. 77. - P. 2255.

69.Tokito S. Metal Oxides as a Hole-Injecting Layer for an Organic Electroluminescent Device / S. Tokito, K. Noda and Y. Taga // J. Phys. D. - 1996. - V. 29. - P. 2750-2753.

70. Zhou X. Very-Low-Operating-Voltage Organic Light-Emitting Diodes Using a p-Doped Amorphous Hole Injection Layer / X. Zhou, M. Pfeiffer, J. Blochwitz [et al.] Phys. Lett. - 2001. - V. 78. - P. 410.

71. Murase S. Solution Processed MoO_3 Interfacial Layer for Organic Photovoltaics Prepared by a Facile Synthesis Method / S. Murase, Y. Yang // Adv. Mater. - 2012. - V. 24. - P. 2459-2462.

72. Inigo A. R. Carbon Nanotube Modified Electrodes for Enhanced Brightness in Organic Light Emitting Devices / A. R. Inigo, J. M. Underwood, S. R. P. Silva // Carbon. - 2011. - V. 49. - P. 4211-4217.

73. Tan L. W. Modification of Charge Transport in Triphenyldiamine Films Induced by Acid Oxidized Single-Walled Carbon Nanotube Interlayers / L. W. Tan, R. A. Hatton, G. Latini [et al.] 19. - P. 485706.

74. Hatton R. A. Carbon Nanotubes: a Multi-Functional Material for Organic Optoelectronics / R. A. Hatton, A. J. Miller and S. R. P. Silva // J. Mater. Chem. - 2008. - V. 18. - P. 1183-1192.

75. Han T. H. Extremely Efficient Flexible Organic Light-Emitting Diodes with Modified Graphene Anode / T. H. Han, Y. Lee, M. R. Choi [et al.] Photonics. - 2012. - V. 6. - P. 105-110.

76. Wu J. Organic Light-Emitting Diodes on Solution-Processed Graphene Transparent Electrodes / J. Wu, M. Agrawal, H. A. Becerril [et al.] - 2010. - V. 4. - P. 43-48.

77. Sun T. Multilayered Graphene Used as Anode of Organic Light Emitting Devices / T. Sun, Z. L. Wang, Z. J. Shi [et al.] Phys. Lett. - 2010. - V. 96. - P. 133301.

78. Zhong Z. Facile Synthesis of Organo-Soluble Surface-Grafted All-Single-Layer Graphene Oxide as Hole-Injecting Buffer Material in Organic Light-Emitting Diodes / Z. Zhong, Y. Dai, D. Ma and Z. Wang // J. Mater. Chem. - 2011. - V. 21. - P. 6040-6045.

79.Lee D. W. Highly Controlled Transparent and Conducting Thin Films Using Layer-by-Layer Assembly of Oppositely Charged Reduced Graphene Oxides / D. W. Lee, T. K. Hong, D. Kang [et al.] Chem. - 2011. - V. 21. - P. 3438-3442.

80.Hwang J. O. Workfunction-Tunable, N-Doped Reduced Graphene Transparent Electrodes for High-Performance Polymer Light-Emitting Diodes / J. O. Hwang, J. S. Park, D. S. Choi [et al.] - 2012. - V. 6. - P. 159-167.

81.Matyba P. Graphene and Mobile Ions: The Key to All-Plastic, Solution-Processed Light-Emitting Devices / P. Matyba, H. Yamaguchi, G. Eda [et al.] - 2010. - V. 4. - P. 637-642.

82.Lee B. R. Highly Efficient Polymer Light-Emitting Diodes Using Graphene Oxide as a Hole Transport Layer / B. R. Lee, J. Kim, D. Kang [et al.] - 2012. - V. 6. - P. 2984-2991.

83.Hrytsak (Toporovska) L. R. Synthesis and characterization of materials with different dimensions based on ZnO [Text] : dissertation for the degree of Doctor of Philosophy in specialty 105 "Applied Physics and Nanomaterials" (10 - Natural Sciences) : defended on 22.05.2021 : approved on 29.06.2021 / Hrytsak (Toporovska) Lilia Romanivna - Ivan Franko National University of Lviv, Lviv, 2021. 161 p.

84.Savchak M. Highly Conductive and Transparent Reduced Graphene Oxide Nanoscale Films via Thermal Conversion of Polymer-Encapsulated Graphene Oxide Sheets / M. Savchak, N. Borodinov, R. Burtovyy [et al.] 10. - P. 3975-3985.

85.Strouk A. L. Photochemical Reduction of Graphene Oxide in a Colloidal Solution / A. L. Strouk, N. S. Andryushin, N. D. Scherban [et al. // Theoretical and Experimental Chemistry. - 2012. - V. 48. - P. 1-10.

86. Eda G. Chemically Derived Graphene Oxide Towards Large-Area Thin-Film Electronics and Optoelectronics / G. Eda, M. Chhowalla // Advanced Material Journal. - 2010. - V. 22. - P. 2392-2415.

87. Zhai J. Visible-Light Photocatalytic Activity of Graphene Oxide-Wrapped Bi WO_{26} Hierachical Microspheres / J. Zhai, H. Yu, H. Li [et al.] 344. - P. 101-106.

88. Wang Y. Y. Crystal Growth, Structure and Optical Properties of Solvated Crystalline Tris(8-Hydroxyquinoline)Aluminum (III) (Alq_3) / Y. Y. Wang, Y. Ren, J. Liu [et al. // Dyes and Pigments. - 2016. - V. 133. - P. 9-15.

89. Karbovnyk I. Polarized Photoluminescence of Alq_3 Thin Films Obtained by the Method of Oblique-Angle Deposition / I. Karbovnyk, B. Sadovyi, B. Turko [et al.] Opt. - 2021. - V. 22. - P. 209-215.

90. Karbovnyk I. Optical Properties of Composite Structure Based on ZnO Microneedles and Alq_3 Thin Film / I. Karbovnyk, B. Sadovyi, B. Turko [et al.] - 2021. - V. 53. - P. 647 (9 pp).

91. Turko B. I. Investigation of the Intrinsic Absorption Edge in Nanostructured Polycrystalline Zinc Oxide Thin Films / B. I. Turko, V. B. Kapustianyk, V. P. Rudyk [et al.] -2006. - V. 73.   P. 222-226.

92. Turko B. I. Manifestation of a Size Effect in the Behavior of the Intrinsic Absorption Edge of Nanostructured Polycrystalline Zinc Oxide Thin Films / B. I. Turko, V. B. Kapustyanyk, V. P. Rudyk [et al.] -2007. - V. 74. P. 310-312.

93. Kapustianyk V. B. Effect of Dopants and Surface Morphology on the Absorption Edge of ZnO Films Doped with In, Al, and Ga // V. B. Kapustianyk, B. I. Turko, V. P. Rudyk [et al.] 82. - P. 153-156.

94. Zhong Z. Y. Optical and Electrical Properties of Some Optical Thin Film Materials / Z. Y. Zhong, Y. D. Jiang, W.Z. Li, X. Yang // Proc. SPIE. - 2006. - V. 6149. - P. 61492E.

95. El-Nahass M. M. Structural and Optical Properties of Tris(8-Hydroxyquinoline) Aluminum (III) (Alq$_3$) Thermal Evaporated Thin Films / M. M. El-Nahass, A. M. Farid, A. A. Atta // Journal of Alloys and Compounds. - 2010. - V. 507. - P. 112-119.

96. Muhammad F. F. Utilizing a Simple and Reliable Method to Investigate the Optical Functions of Small Molecular Organic Films - Alq$_3$ and Gaq$_3$ as Examples / F. F. Muhammad, K. Sulaiman // Measurement. - 2011. - V. 44. - P. 1468-1474.

97. Organometallic Luminescence: A Case Study on Alq$_3$, an OLED Reference Material / G. Baldacchini - Cambridge: Woodhead Publishing, 2020. - 352 p.

98. Brinkmann M. Correlation between Molecular Packing and Optical Properties in Different Crystalline Polymorphs and Amorphous Thin Films of mer-Tris(8-Hydroxyquinoline)Aluminum (III) / M. Brinkmann, G. Gadret, M. Muccini [et al.] Chem. Soc. - 2000. - V. 122. - P. 5147-5157.

99. Baldacchini G. Optical Spectroscopy of Tris(8-Hydroxyquinoline) Aluminum Thin Films / G. Baldacchini, S. Gagliardi, R. M. Montereali [et al.] 82. - P. 669-680.

100. Colle M. Delayed Fluorescence and Phosphorescence of Tris-(8-Hydroxyquinoline)Aluminum (Alq$_3$) and Their Temperature Dependence / M. Colle, C. Garditz // Journal of Luminescence. - 2004. - V. 110. - P. 200-206.

101. Colle M. The Triplet State in Tris-(8-Hydroxyquinoline)Aluminum / M. Colle, C. Garditz, M. Braun // Journal of Applied Physics. - 2004. - V. 96. - P. 6133-6141.

102. M. Colle and W. Brutting. Chapter 4. Thermal and Structural Properties of the Organic Semiconductor Alq$_3$ and Characterization of Its Excited Electronic Triplet State. In Book: Physics of Organic Semiconductors / Ed. by W. Brutting. - Weinheim: Wiley-VCH Verlag GmbH & Co. KGaA, 2005. - 536 p.

103. Garbuzov D. Z. Photoluminescence Efficiency and Absorption of Aluminum-Tris-Quinolate (Alq$_3$) Thin Films / D. Z. Garbuzov, V. Bulovic, P. E. Burrows, S. R. Forrest // Chemical Physics Letters. - 1996. - V. 249. - P. 433-437.

104. Aziz A. Optical Absorption in AlQ/A. Aziz, K. L. Narasimhan // Synthetic Metals. - 2000. - V. 114. - P. 133-137.

105. Ravi Kishore V. V. N. On the Assignment of the Absorption Bands in the Optical Spectrum of Alq$_3$ / V. V. N. Ravi Kishore, A. Aziz, K. L. Narasimhan [et al. // Synthetic Metals. - 2002. - V. 126. - P. 199-205.

106. Muhammad F. F. Study of Optoelectronic Energy Bands and Molecular Energy Levels of Tris (8-Hydroxyquinolinate) Gallium and Aluminum Organometallic Materials from Their Spectroscopic and Electrochemical Analysis / F. F. Muhammad, A. I. A. Hapip, K. Sulaiman // Journal of Organometallic Chemistry. - 2010. - V. 695. - P. 2526-2531.

107. Cuong N. K. Study on Current-Voltage Characteristics of OLEDs Using Alq$_3$ as the Electron Transport Layer / N. K. Cuong // VNU Journal of Science, Mathermatics - Physics. - 2011. - V. 27. - P. 174-180.

108. Makki A. H. Yellow Emissive Tris(8-Hydroxyquinoline) Aluminum by the Incorporation of ZnO Quantum Dots for OLED Applications / A. H. Makki, S.-H. Park // Micromachines. - 2021. - V. 12. - P. 1173.

109. Xie W. Structural and Optical Properties of ε-Phase Tris(8-Hydroxyquinoline) Aluminum Crystals Prepared by Using Physical Vapor Deposition Method / W. Xie, Z. Pang, Y. Zhao [et al. // Journal of Crystal Growth. - 2014. - V. 404. - P. 164-167.

110. Ajward A. M. Radiative Recombination of Trapped Excitons in Alq_3 Films: Importance of Intermolecular Interactions / A. M. Ajward, X. Wang, N. Wickremasinghe [et al. // Physical Review B. - 2013. - V. 88. - P. 045205.

111. Walser A. D. Temperature Dependence of the Singlet Excited State Lifetime in Alq_3 / A. D. Walser, R. Priestley, R. Dorsinville // Synth. Met. - 1999. - V. 102. - P. 1552-1553.

112. Saha S. K. Temperature- and Field-Dependent Quantum Efficiency in Tris-(8-Hydroxy) Quinoline Aluminum Light-Emitting Diodes / S. K. Saha, Y. K. Su, F. S. Juang // J. Appl. Phys. - 2001. - V. 89. - P. 8175-8178.

113. Kaiser C. A Universal Urbach Rule for Disordered Organic Semiconductors / C. Kaiser, O. J. Sandberg, N. Zarrabi [et al.] 12. - P. 3988.

114. Muhammad F. F. Optical Response and Photovoltaic Performance of Organic Solar Cells Based on $DH6T$:Alq_3 Active Layer / F. F. Muhammad, A. J. Muhammad, K. Sulaiman // Journal of Technology Innovations in Renewable Energy. - 2016. - V. 5. - P. 3-10.

115. Chen C.-T. Evolution of Red Organic Light-Emitting Diodes: Materials and Devices / C.-T. Chen // Chem. Commun. - 2004. - V. 16. - P. 4389-4400.

116. Guo Z. Dicyanomethylene-4H-Pyran Chromophores for OLED Emitters, Logic Gates and Optical Chemosensors / Z. Guo, W. Zhu and H. Tian // Chem. Commun. - 2012. - V. 48. - P. 6073-6084.

117. Photophysical Properties of 4-(Dicyanomethylene)-2-Methyl-6-(4-Dimethylaminostyryl)-4H-Pyran (DCM) and Optical Sensing Applications. Chapter 1 in Book: Photophysics, Photochemical and Substitution Reactions - Recent Advances / ed. by S. Saha, R. K. Kanaparthi, T. Soldatovic: TechOpen CY, 2021. - 230 p.

118. Wang C. Trace Detection of RDX, HMX and PETN Explosives Using a Fluorescence Spot Sensor / C. Wang, H. Huang, B. R. Bunes [et al.] 6. - P. 25015 (9 pp).

119. El-Shishtawy R. M. Pyran-Squaraine as Photosensitizers for Dye-Sensitized Solar Cells: DFT/TDDFT Study of the Electronic Structures and Absorption Properties / R. M. El-Shishtawy, S. A. Elroby, A. M. Asiri, R. H. Hilal // International Journal of Photoenergy. - 2014. - V. 2014. - P. 136893 (11 pp).

120. Tang C. Electroluminescence of Doped Organic Thin Films / C. Tang, S. VanSlyke // Journal of Applied Physics. - 1989. - V. 65. - P. 3610-3616.

121. Yao, Y.-S. Starburst DCM-Type Red-Light-Emitting Materials for Electroluminescence / Y.-S. Yao, J. Xiao, X.-S. Wang [et al. // Applications Advanced Functional Materials. - 2006. - V. 16. - P. 709-718.

122. Salehi A. Manipulating Refractive Index in Organic Light-Emitting Diodes / A. Salehi, Y. Chen, X. Fu, C. Peng, F. So // ACS Appl. Mater. Interfaces. - 2018. - V. 10. - P. 9595-9601.

123. Szeto B. Obliquely Deposited tris(8-Hydroxyquinoline) Aluminum (Alq$_3$) Biaxial Thin Films with Negative in-Plane Birefringence / B. Szeto, P. C. P. Hrudey, J. Gospodyn [et al.] A Pure Appl. Opt. - 2007. - V. 9. - P. 457-462.

124. Gromov V.K. Introduction to ellipsometry / V.K. Gromov - Leningrad: Leningrad University Press, 1986. 190 p.

125. Bazyl O. K. Features of the Electronic Structure and Photophysical Processes in Asymmetric and Symmetric (Dicyanomethylene)-Pyran Dyes / O. K. Bazyl, V. A. Svetlichnyi // Opt. Spectrosc. - 2015. - V. 118. - P. 37-45.

126. Karbovnik I. Polarized photoluminescence of thin films of dicyanomethylene pyran and its derivatives obtained by oblique angle deposition / I. Karbovnyk. Karbovnyk, B. Turko, V. Vasyliev [et al. Physical Series. - 2021. - Vol. 58. - pp. 50-60.

127. Zaki T. Short-Channel Organic Thin-Film Transistors: Fabrication, Characterization, Modeling and Circuit Demonstration / T. Zaki: Springer, 2015. - 220 p.

128. Barranco A. Perspectives on Oblique Angle Deposition of Thin Films: From Fundamentals to Devices / A. Barranco, A. Borras, A. R. Gonzalez-Elipe, A. Palmero // Prog. Mater. Sci. - 2016. - V. 76. - P. 59-153.

129. Zhang D.-W. Recent Advances in Circularly Polarized Electroluminescence Based on Organic Light-Emitting Diodes / D.-W. Zhang, M. Li and C.-F. Chen // Chem. Soc. Rev. - 2020. - V. 49. - P. 1331-1343.

130. Hrudey P. C. P. Highly Ordered Organic Alq_3 Chiral Luminescent Thin Films Fabricated by Glancing-Angle Deposition / P. C. P. Hrudey, K. L. Westra, M. J. Brett // Adv. Mater. - 2006. - V. 18. - P. 224-228.

131. Muccini M. Blue Luminescence of Facial Tris(Quinolin-8-Olato)Aluminum (III) in Solution, Crystals, and Thin Films / M. Muccini, M. A. Loi, K. Kenevey [et al. // Adv. Mater. - 2004. - V. 16 - P. 851-888.

132. Rajeswaran M. Single-Crystal Structure Determination of a New Polymorph (ε-Alq_3) of the Electroluminescence OLED (Organic Light-Emitting Diode) Material, Tris(8-Hydroxyquinoline)Aluminum (Alq_3) / M. Rajeswaran, T. N. Blanton // Journal of Chemical Crystallography. - 2005. - V. 35. - P. 71-76.

133. Cui C. Oligonucleotide Assisted Light-Emitting Alq_3 Microrods: Energy Transfer Effect with Fluorescent Dyes / C. Cui, D. H. Park, J. Kim [et al.] Commun. - 2013. - V. 49. - P. 5360-5362.

134. Park J. Fine Fabrication and Optical Waveguide Characteristics of Hexagonal Tris(8-Hydroxyquinoline)Aluminum (III) (Alq_3) Crystal / J. Park, S. Kim, J. Choi [et al.] 10. - P. 260 (8 pp).

135. Braun M. A New Crystalline Phase of the Electroluminescent Material Tris(8-Hydroxyquinoline) Aluminum Exhibiting Blueshifted Fluorescence / M. Braun, J. Gmeiner, M. Tzolov [et al.] Phys. - 2001. - V. 114. - P. 9625-9632.

136. Rajeswaran M. Structural, Thermal, and Spectral Characterization of the Different Crystalline Forms of Alq_3 , Tris(Quinolin-8-Olato)Aluminum (III), an Electroluminescent Material in OLED Technology / M. Rajeswaran, T. N. Blanton, C. W. Tang [et al.] 28. - P. 835-843.

137. Neumann A. Opto-Valleytronic Imaging of Atomically Thin Semiconductors / A. Neumann, J. Lindlau, L. Colombier [et al.] 12. - P. 329-334.

138. Okabayashi Y. Positive Giant Surface Potential of Tris(8-Hydroxyquinolinolato) Aluminum (Alq_3) Film Evaporated onto Backside of Alq_3 Film Showing Negative Giant Surface Potential / Y. Okabayashi, E. Ito, T. Isoshima, M. Har // Applied Physics Express. - 2012. - V. 5. - P. 055601.

139. Marchetti A. P. Permanent Polarization and Charge Distribution in Organic Light-Emitting Diodes (OLEDs): Insights from Near-Infrared Charge-Modulation Spectroscopy of an Operating OLED / A. P. Marchetti, T. L. Haskins, R. H. Young, L. J. Rothberg // J. Appl. Phys. - 2014. - V. 115. - P. 114506.

140. Ito E. Spontaneous Buildup of Giant Surface Potential by Vacuum Deposition of Alq_3 and Its Removal by Visible Light Irradiation / E. Ito, Y. Washizu, N. Hayashi [et al.] Phys. - 2002. - V. 92. - P. 7306-7310.

141. Kajimoto N. Decay Process of a Large Surface Potential of Alq_3 Films by Heating / N. Kajimoto, T. Manaka, M. Iwamotoa // J. Appl. Phys. - 2006. - V. 100. - P. 053707.

142. Sugi K. Characterization of Light-Erasable Giant Surface Potential Built up in Evaporated Alq_3 Thin Films / K. Sugi, H. Ishii, Y. Kimura [et al. // Thin Solid Films. - 2004. - V. 464-465. - P. 412- 415.

143. Yoshizaki K. Large Surface Potential of Alq_3 Film and Its Decay / K. Yoshizaki, T. Manaka, M. Iwamoto // J. Appl. Phys. - 2005. - V. 97. - P. 023703.

144. Noguchi Y. Threshold Voltage Shift and Formation of Charge Traps Induced by Light Irradiation During the Fabrication of Organic Light-Emitting Diodes / Y. Noguchi, N. Sato, Y. Tanaka [et al.] Phys. Lett. - 2008. - V. 92. - P. 203306.

145. Isoshima T. Long-Term Relaxation of Molecular Orientation in Vacuum-Deposited ALQ_3 Thin Films / T. Isoshima, H. Ito, E. Ito [et al.] Cryst. Liq. Cryst. - 2009. - V. 505. - P. 59/[297]-63/[301].

146. Datta K. Strain Sensitivity of Dielectric Polarization to Doping in a Host: Guest Medium / K. Datta, P. B Deotare // Optical Materials Express. - 2020. - V. 10. -- P. 3021-3029.

147. Lee S. Twisted Intramolecular Charge Transfer State of a "Push-Pull" Emitter / S. Lee, M. Jen, Y. Pang // Int. J. Mol. Sci. - 2020. - V. 21. - P. 7999 (15 pp).

148. di Nunzio M. R. Confinement Effect of Micro- and Mesoporous Materials on the Spectroscopy and Dynamics of a Stilbene Derivative Dye / M. R. di Nunzio, G. Perenlei, A. Douhal // Int. J. Mol. Sci. - 2019. - V. 20. - P. 1316 (19 pp).

149. Bondarev S. Fluorescence and Electronic Structure of the Laser Dye DCM in Solutions and in Polymethylmethacrylate / S. Bondarev, V. Knyukshto, V. Stepuro [et al.] 71. - P. 194-201.

150. Tang C. W. Organic Electroluminescent Diodes / C. W. Tang, S. A. Van Slyke // Applied Physics Letters. - 1987. - V. 51. - P. 913-915.

151. Bradley D. Electroluminescent Polymers: Materials, Physics and Device Engineering / D. Bradley // Current Opinion in Solid State and Materials Science. - 1996. - V. 1. - P. 789-797.

152. Grell M. Polarized Luminescence from Oriented Molecular Materials / M. Grell, D. D. Bradley // Advanced Materials. - 1999. - V. 11. - P. 895-905.

153. Fujiki M. Questions of Mirror Symmetry at the Photoexcited and Ground States of Non-Rigid Luminophores Raised by Circularly Polarized Luminescence and Circular Dichroism Spectroscopy. Part 2: Perylenes, BODIPYs, Molecular Scintillators, Coumarins, Rhodamine B, and DCM / M. Fujiki, J. R. Koe, S. Amazumi // Symmetry. - 2019. - V. 11. - P. 363 (41 pp).

154. Karbovnyk I. Formation of Oriented Luminescent Organic Thin Films on Modified Polymer Substrate / I. Karbovnyk, B. Sadovyi, B. Turko [et al.] 10. - P. 2791-2796.

155. Kukhta A. V. Alignment of Luminescent Liquid Crystalline Molecules on Modified PEDOT:PSS Substrate / A. V. Kukhta, S. A. Maksimenko, K. M. Degtyarenko [et al. // Applied Nanoscience. - 2020. - V. 10. - P. 5063-5068.

156. Vanjinathan M. Design, Synthesis, Photophysical, and Electrochemical Properties of DCM-Based Conjugated Polymers for Light-Emitting Devices / M. Vanjinathan, H.-C. Lin, A. S. Nasar // Journal of Polymer Science Part A: Polymer Chemistry. - 2012. - V. 50. - P. 3806-3818.

157. Vembris A. Stimulated Emission and Optical Properties of Pyranylidene Fragment Containing Compounds in PVK Matrix / A.

Vembris, E. Zarins, V. Kokars // Optics & Laser Technology. - 2017. - V. 95. - P. 74-80.

158. Yin Y. Evolution of White Organic Light-Emitting Devices: from Academic Research to Lighting and Display Applications / Y. Yin, M.U. Ali, W. Xie [et al.] Chem. Front. - 2019. - V. 3. - P. 970-1031.

159. Kapustianyk V. LEDs Based on p-Type ZnO Nanowires Synthesized by Electrochemical Deposition Method / V. Kapustianyk, B. Turko, I. Luzinov [et al.] Status Solidi C. - 2014. - V. 11. - P. 1501-1504.

160. Turko B. Electroluminescence from n-ZnO Microdisks/p-GaN Heterostructure / B. Turko, A. Nikolenko, B. Sadovyi [et al.] - 2019. - V. 51. - P. 135 (11 pp).

161. Cuba M. Synthesis and Optical Properties of ZnO Incorporated Tris-(8-Hydroxyquinoline)Aluminum / M. Cuba, U. Rathinavalli, K. Thangaraju, G. Muralidharan // Journal of Luminescence. - 2014. - V. 153. - P. 188-193.

162. Cuba M. Enhanced Luminescence Properties of Hybrid Alq_3 /ZnO (Organic/Inorganic) Composite Films / M. Cuba, G. Muralidharan // Journal of Luminescence. - 2014. - V. 156. - P. 1-7.

163. Dasi G. Improved Electron Injection in Spin Coated Alq_3 Incorporated ZnO Thin Film in the Device for Solution Processed OLEDs / G. Dasi, R. Ramarajan and K. Thangaraju // Journal of Applied Physics. Conference Proceedings. - 2018. - V. 1942. - P. 060015.

164. Sanchez-Valencia J. R. Growth Assisted by Glancing Angle Deposition: a New Technique to Fabricate Highly Porous Anisotropic Thin Films / J. R. Sanchez-Valencia, R. Longtin, M. D. Rossell, P. Gröning // ACS Applied Materials and Interfaces. - 2016. - V. 8. - P. 8686-8693.

165. Iechi H. Vertical Type Organic Light Emitting Transistor Using Thin-Film ZnO/H. Iechi, M. Sakai, M. Nakamura, K. Kudo // Book of abstracts

International conference on solid state devices and materials, September 15-17, 2004. - Tokyo, Japan. - P. 164-165.

166. Kan P. Electroluminescence Dependence on the Organic Thickness in ZnO Nano Rods/Alq$_3$ Heterostructure Devices / P. Kan, Y. Wang, S. Zhao [et al. // Journal of Nanoscience and Nanotechnology. - 2011. - V. 11. - P. 3470-3473.

167. Turko B. I. Ultraviolet Electroluminescence of LED Devices Based on n-ZnO Nanorods Grown by Various Methods and p-GaN Films / B. I. Turko, A. S. Nikolenko, B. S. Sadovyi [et al.] 25. - P. 1701 (6 pp).

168. Kapustianyk V. Effect of Vacuumization on the Photoluminescence and Photoresponse Decay of the Zinc Oxide Nanostructures Grown by Different Methods / V. Kapustianyk, B. Turko, V. Rudyk [et al.] 56. - P. 71-74.

169. Kapustianyk V. Exciton Spectra of the Nanostructured Zinc Oxide / V. Kapustianyk, M. Panasiuk, G. Lubochkova [et al.] 12. - P. 2602.

170. Cao H. Microlaser Made of Disordered Media / H. Cao, J. Xu, E. Seelig, R. Chang // Appl. Phys. Letters. - 2000. - V. 76. - P. 2997-2999.

171. Tang Z. Room-Temperature Ultraviolet Laser Emission from Self-Assembled ZnO Microcrystallite Thin Films / Z. Tang, G. Wong, P. Yu [et al.] Phys. Letters. - 1998. - V. 72. - P. 3270-3272.

172. Gruzintsev A. Edge Luminescence of ZnO Nanorods on High-Intensity Optical Excitation / A. Gruzintsev, A. Redkin, E. Yakimov, C. Barthou // Semiconductors. - 2008. - V. 42. - P. 1092-1097.

173. Yase K. Anisotropic Photoluminescence from Alq$_3$ and TPD Films on Solid Substrates / K. Yase, N. Takada, W. Yoshida [et al.] Crysr. Liq. Crysr. - 1996. - V. 280. - P. 379-384.

174. Mori T. Electroluminescence of Organic Light Emitting Diodes with Alternatively Deposited Dye-Doped Aluminum Quinoline and Diamine

Derivative / T. Mori, K. Obata, T. Mizutani // J. Phys. D: Appl. Phys. 1999. - V. 32. - P. 1198-1203.

175. Qin Y. Luminous Composite Ultrathin Films of the DCM Dye Assembled with Layered Double Hydroxides and Its Fluorescence Solvatochromism Properties for Polarity Sensors / Y. Qin, P. Zhang, L. Lai [et al.] 3. - P. 5246-5252.

176. Yokoyama S. Amplified Spontaneous Emission and Laser Emission from a High Optical-Gain Medium of Dyedoped Dendrimer / S. Yokoyama, T. Nakahama, S. Mashiko // Journal of Luminescence. - 2005. - V. 111. - P. 285-290.

177. Zhong G. Y. In Situ Photoluminescence Investigation of Doped Alq / G. Y. Zhong, J. He, S.T. Zhang [et al. // Applied Physics Letters. - 2002. - V. 80. - P. 4846.

178. Punke M. Coupling of Organic Semiconductor Amplified Spontaneous Emission Into Polymeric Single-Mode Waveguides Patterned by Deep-UV Irradiation / M. Punke, S. Mozer, M. Stroisch [et al.] 19. - P. 61-63.

179. Kaur A. Voltage Tunable Multicolor Light Emitting Diodes Based on a Dye-Doped Polythiophene Derivative / A. Kaur, M. J. Cazeca, S. K. Sengupta [et al. // Synthetic Metals. - 2002. - V. 126. - P. 283-288.

180. Kwak H. S. Design of Organic Electronic Materials with a Goal-Directed Generative Model Powered by Deep Neural Networks and High-Throughput Molecular Simulations / H. S. Kwak, Y. An, D. J. Giesen [et al.] 9. - P. 800370.

181. Miao Y. Recent Progress on Organic Light-Emitting Diodes with Phosphorescent Ultrathin (<1 nm) Light-Emitting Layers / Y. Miao, M. Yin // iScience. - 2022. - V. 25. - P. 103804.

182. Su R. 3D-Printed Flexible Organic Light-Emitting Diode Displays / R. Su, S.H. Park [et al.] Adv. - 2022. - V. 8. - P. eabl8798.